# Contents

# Future of Denial

## *The Ideologies of Climate Change*

Tad DeLay

**VERSO**
London • New York

For Logan and Asher

First published by Verso 2024
© Tad DeLay 2024

1 3 5 7 9 10 8 6 4 2

**Verso**
UK: 6 Meard Street, London W1F 0EG
US: 388 Atlantic Avenue, Brooklyn, NY 11217
versobooks.com

Verso is the imprint of New Left Books

ISBN-13: 978-1-83976-543-8
ISBN-13: 978-1-83976-549-0 (US EBK)
ISBN-13: 978-1-83976-548-3 (UK EBK)

**British Library Cataloguing in Publication Data**
A catalogue record for this book is available from the British Library

**Library of Congress Cataloging-in-Publication Data**

Names: DeLay, Tad, author.
Title: Future of denial : the ideologies of climate change / Tad DeLay.
Description: New York : Verso, 2024. | Includes bibliographical references
   and index.
Identifiers: LCCN 2023046425 (print) | LCCN 2023046426 (ebook) | ISBN
   9781839765438 (hardback) | ISBN 9781839765490 (ebook)
Subjects: LCSH: Climatic changes—Psychological aspects. | Environmental
   psychology.
Classification: LCC BF353.5.C55 D45 2024  (print) | LCC BF353.5.C55
   (ebook) | DDC 155.9/15—dc23/eng/20231106
LC record available at https://lccn.loc.gov/2023046425
LC ebook record available at https://lccn.loc.gov/2023046426

Typeset in Sabon by MJ & N Gavan, Truro, Cornwall
Printed and bound by CPI Group (UK) Ltd, Croydon, CR0 4YY

# List of Figures

# Introduction

"You're not from around here, are you?"

It is autumn 2020. Wildfires stretching the West Coast paint the skies burnt orange. In various places the hue at noon is blood red. Five of the ten largest wildfires in California history burn simultaneously.

Nobody alive has witnessed such ruin. It has not been so hot since the Eemian interglacial period some 125,000 years ago, when temperatures averaged a degree warmer, testing our ancestors' abilities to adapt while seas rose 130 meters over fifteen millennia. More recently, in the Bølling-Allerød interstadial 14,700 years ago, toward the end of the last glacial period, the sea level rose four or five meters per century for five centuries. As the Laurentide Ice Sheet covering much of North America retreated, the Great Lakes and the massive Lake Agassiz settled until the latter emptied its freshwater into the Atlantic and disrupted the overturning of cold and warm water. The resulting thousand-year cold snap, the Younger Dryas, ended abruptly 11,700 years ago when temperatures rose several degrees in a few decades. All this came before the Holocene. Between the origins of agriculture in the Fertile Crescent and the invention of hydraulic fracturing, indeed in the entire history of civilization after we left the cave, Earth has never felt so hot, never changed so fast. But paleoclimate data is not on the mind of the vigilante barking, "You're not from around here, are you?"

He is armed. He points his question at an African American mother fleeing wildfires. She registers the racial overtones of his inquisition.

She's stopped at an improvised checkpoint in Corbett, Oregon. Fires threaten communities around Portland, a white area of the Pacific Northwest. Evacuees take to the roads.

It is the year of the pandemic, just after a summer of mass protests for racial justice in the wake of murders by police. An election approaches. The incumbent praises violence. Police assault countless citizens. Over a hundred reactionaries ram protestors with vehicles; conservative legislatures rush through bills to authorize the manslaughter tactic. Hundreds of thousands die of a virus. America is about to erupt, wallowing in a desire for . . . what? Not a desire to know. You're not from around here, are you?

His eyes dart from her to the other occupants. Scan the vehicle. Gauge threats. In the back, her young children stare back at the vigilante and the weapon with which he might dispatch them.

No, these are not the targets. He is searching for arsonists, saboteurs, the right's hallucinated fixation: antifa.

In nearby Estacada, a journalist is accosted and flees. As he speeds away, a truck intercepts from the front and blocks the highway. Out of the vehicle emerges a gunman aiming his rifle through the journalist's windshield.

A few miles southwest in Molalla, three journalists are stopped at another improvised checkpoint, guns raised. A vigilante snaps, "Get the fuck out of here," as he takes photos of their faces and license plate.

"It seems like the militants are burning out/up the rural folks closest to the cities," says one member of a Facebook group organizing the counterattack, "because the only way they can fight is dirty vs the more conservative/rural folks would hand them their @$$es [sic] in an altercation." Another suggests they fight. "Most of us can mobilize and bring our arsenal with us." The radio reports antifa shooting at firefighters.

Police are alerted to illegal checkpoints but do not intervene. After a school board member organizes citizen patrols, a sheriff's sergeant asks for "photos of cars and even license plates." He advises vigilantes keep an eye out for "anything that feels out of place to you, just listen to your gut because nine times out of ten your gut is right."

A Clackamas County sheriff's deputy is forced out of his job after a video goes viral in which he erupts at the imaginary culprits. "What I'm worried about is that there's people stashing stuff. It means that they're gonna go in preparation. And I don't wanna sound like some doomsdayer, but it's getting serious, and, I, we need the public's help on this—"

Another interjects to say people around Portland will simply stay home and die in the inferno rather than risk evacuating only to encounter anarchists. The deputy blurts, "Antifa motherfuckers, okay, are out causing hell. And there's a lot of lives at stake, and there's a lot of people's property at stake, because these guys got some vendetta."

The next day, this same deputy advises a citizen's group that they must be cautious when using deadly force. When they kill outsiders, he tells them, the courts will demand explanation. He laughs with them, charms the bloodthirsty piglets. "Now, you throw a fucking knife in their hand after you shoot 'em, that's on you." More laughter from the crowd. He assures them, "I am on your side, people, I am one hundred percent! I wouldn't let this shit happen in my neighborhood either, but be smart about it."

Over a weekend in September 2020, illegal checkpoints and armed squads spring up independently in at least three towns near Portland. Fueled on social media rumors and encouraged by police, citizens organically convert latent climate denial, scouting instead for imaginary leftists while pointing weapons at reporters and families. Thankfully, there are no reports of lethal violence this time. A near miss. You're not from around here, are you?

The age of denial is over, or so we are regularly told. With each new climate report, protest, and catastrophe, we hear "the age of denial is over," often with that precise phrasing. More familiar words and phrases in climate journalism include: the Anthropocene; evidence is now unequivocal; made worse by climate change; the effects of which will fall disproportionately on younger generations; call for action at the next summit; net-zero emissions by 2050; 97 percent of scientists agree; window for action is closing. You could layer clichés and build a template for nine-tenths of the next article. The age of denial is over, and yet!

Americans use Celsius only to talk about the end of the world. In 2018 the Intergovernmental Panel on Climate Change (IPCC) published a special report showing emissions must plummet by half by 2030 and zero by 2050 to have a chance at limiting warming to 1.5°C. In those twelve years, we—so much work is done, so many interests laundered, by this little pronoun—must decouple capitalism from carbon. In an interview, newly elected Representative Alexandria Ocasio-Cortez summarized, "Millennials and Gen Z and all these folks that come after us are looking up, and we're like, 'The world is going to end in twelve years if we don't address climate change, and your biggest issue is how are we gonna pay for it?'"

The twelve-year deadline briefly turned into an activist's cry. But to those cackling at the thought of ecocide, the deadline was hyperbolic whining. Ocasio-Cortez's phrase "the world is going to end" gave them a chance to feign ignorance.

Three years after this exchange, the International Energy Agency (IEA) updated its forecasts for energy and emissions after the Covid-19 recovery. It was the same year Jeff Bezos and Richard Branson launched themselves into space on rockets emitting as much per passenger as anyone from the poorest billion emits in an entire lifetime.[1] Governments weren't doing enough to transition, said the IEA, weren't investing where needed. The notoriously conservative organization's president called for a ban on new fossil fuel infrastructure: "If governments are serious about the climate crisis, there can be no new investments in oil, gas and coal, from now—from this year."[2]

Merely 2 percent of pandemic recovery funds went to energy-related sustainability. A few nations were indeed reducing emissions, the United States among them, good news in cases of authentic decrease but less so when drops are paper-thin accounting tricks balanced on outsourced production. At any rate, the fall wasn't enough. IEA forecasts show emissions setting new records in 2023, followed by "no clear peak in sight."[3]

Annually, our mode of production pumps out approximately forty-one gigatonnes of carbon dioxide ($41.4 \text{ GtCO}_2$). About 37.5 $\text{GtCO}_2$ come from fossil emissions (electricity, transport, industry, and buildings). The remaining 3.9 $\text{GtCO}_2$ come from agriculture and land use.[4]

---

1   Lucas Chancel et al., "World Inequality Report 2022," World Inequality Lab, 2022, 134.

2   Fiona Harvey, "No New Oil, Gas or Coal Development If World Is to Reach Net Zero by 2050, Says World Energy Body," *The Guardian*, May 18, 2021, theguardian.com.

3   Anmar Frangoul, "$CO_2$ Emissions Set to Hit Record Levels in 2023 and There's 'No Clear Peak in Sight' IEA Says," *CNBC*, July 20, 2021, cnbc.com.

4   Carbon dioxide emissions sources from Pierre Friedlingstein et al., "Global Carbon Budget 2022," *Earth System Science Data* 14, no. 11 (2022): 4811–4900. Global primary energy consumption by source from Hannah Ritchie, Max Roser, and Pablo Rosado "Energy," Our World in Data, 2022, ourworldindata.org.

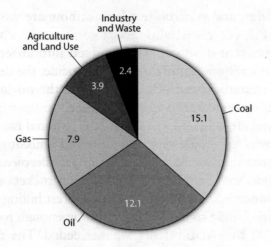

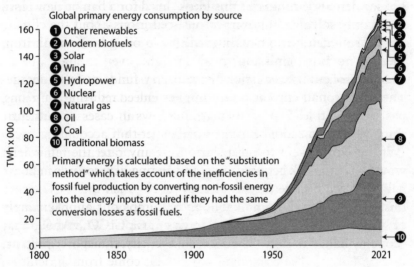

Figure 0.1.  Carbon Dioxide Emissions and Global Primary Energy Consumption by Source

The vast majority of the fossil emissions are from combusted fossil fuels at 35.1 $GtCO_2$. The other 2.4 $GtCO_2$ are landfills and industrial processes such as chemicals and cement, which could still cause emissions even if fossil fuel use ceased. Cement and steel are the largest of this latter group. Burning coal, oil, and natural gas account for eight-tenths of total anthropogenic carbon dioxide emissions.[5]

_____

5   Sources for Figure 0.1: see fossil fuel emissions per Pierre Friedling-stein et al., "Global Carbon Budget 2022." Total emissions are rounded to

Worldwide carbon dioxide emissions and gross domestic product (GPD) correlate tightly (0.93 correlation). Further close correlations include world GDP and primary energy (0.92), atmospheric carbon dioxide and world primary energy (0.98), atmospheric carbon dioxide concentration and world GDP (0.96), atmospheric carbon dioxide concentration and world GDP per capita (0.98), and world GDP growth and world primary energy growth (0.83). Decoupling these relationships isn't quite within our capabilities. At least not yet, not globally. Some misleadingly claim we've already decoupled, such as President Obama when he bragged of economic growth and shrinking emissions, a decline largely due to a transition from one fossil fuel (coal) to another fossil fuel (natural gas, or methane). We will examine similar denials throughout this book, myths in which decoupling is not only solvable but, miraculously, already solved!

Unfortunately, the quants at the IEA are right. As of this writing there is no fossil emissions peak in sight, even though, due to changes in land use, emissions have been nearly flat for a decade. The IPCC found current policy implies emissions continue rising past 2025 and lead to 3.2°C warming by end of century (though some recent projections put the number at 2.8°C).[6]

If the world were to meet Paris Agreement targets, emissions would have to peak between 2020 and 2025 and then fall by almost half within a decade (it was already 2022 when the IPCC wrote this!). Such a drop is unthinkable, given the recent, painful example of what it would mean to wind down emissions. When the global economy halted to slow Covid-19, emissions declined by 5 percent,

---

45 $GtCO_2$ per Intergovernmental Panel on Climate Change, *Sixth Assessment Report*, Working Group III, 2022, 7, but in this case, I have chosen to use Global Carbon Project numbers as they are more precise. See also Hannah Ritchie and Max Roser, "$CO_2$ and Greenhouse Gas Emissions," Our World in Data, May 11, 2020, ourworldindata.org. Note the final 2022 Global Carbon Budget also puts anthropogenic emissions slightly lower at 40.5 GtCO2, partly due to a small concrete carbonation sink, and uncertainties amounting to a few gigatonnes remain. My numbers match the public presentation of the report for Pierre Friedlingstein et al., "Global Carbon Budget 2022," globalcarbonproject.org.

6 Intergovernmental Panel on Climate Change, *Sixth Assessment Report*, Working Group III, 2022, 17.

about 2.5 $GtCO_2$.[7] In the carbon budgets, the lockdowns postponed apocalypse by three weeks.

How bad could it get if, say, our capitalist mode of production incentivized burning through a sizable chunk of fossil fuel reserves worth a couple hundred trillion dollars? The relationship between carbon dioxide and temperature is linear (with uncertainties). Per IPCC, "Each 1,000 $GtCO_2$ of cumulative $CO_2$ emissions is assessed to *likely* cause a 0.27°C to 0.63°C increase in global surface temperature with a best estimate of 0.45°C."[8] So more or less, to figure the potential take the amount of carbon dioxide we can emit, divide by 1,000 and multiply by 0.45.

Combusting all remaining fossil fuel reserves (the portion of resources we can extract with today's technology and prices) would release 3,328 $GtCO_2$. That's enough to raise temperature 1.5°C on top of 1.3°C thus far, bringing us to 2.8°C. It's far from an upper limit for several reasons: inaccessible resources can turn into more reserves for burning, fossil fuels make up only four-fifths of anthropogenic carbon dioxide emissions, carbon dioxide is not the only greenhouse gas, and we might reduce aerosol pollution currently masking temperature rise. Together these factors are why the IPCC estimates 3.2°C by end of century. Worse yet, new technologies, price fluctuations, and government subsidies can turn unrecoverable resources into lucrative reserves. Firing all fossil fuel resources would liberate 45,740 $GtCO_2$. You can do the math.

The upshot is that by the time this book is published, half the twelve-year timeframe will be gone. Fossil emissions continue their climb. Land temperatures rise faster by half than the global average, so those of us on solid ground already experience 1.9°C. Temperatures have climbed 0.18°C per decade on average, though recent evidence suggests that until midcentury the rate may speed up to as much as 0.36°C per decade.[9] Earth sails past the Paris Agreement

---

7 "Emission Reductions from Pandemic Had Unexpected Effects on Atmosphere," NASA Jet Propulsion Laboratory, accessed March 24, 2022, jpl.nasa.gov.

8 Intergovernmental Panel on Climate Change, *Sixth Assessment Report*, Working Group I, 2021, 28.

9 Hansen et al., "Global Warming in the Pipeline," preprint, July 5, 2023, columbia.edu.

goal of 1.5°C by the mid-2030s and approaches the 2°C threshold by the late 2050s. Instead of falling by half, fossil emissions
increase. A strange thing if the age of denial is over.

We are, all of us, in denial about what awaits us in the Changes.[10]
Not just the conservatives. Liberals and social democrats deny as
well. So, too, does the left. Denial is a contingency of repression,
the putting away of an unpleasant idea. When the repressed surfaces, we react with denial. But repression and the return of the
repressed are one and the same. What cannot be confronted converts into strange commitments and violence.

An easy path ignores the polycrisis *tout court*, but denial takes
active paths as well. Sigmund Freud once observed his analysands,
his patients, responded in one of two ways when a repressed idea
surfaced again. Some analysands rejected reality itself, preferring
instead to live out a delusion. Others rejected some moral attribute
in the idea and denied their own fault. A simple schema: reality
denial and guilt denial.

When it comes to climate change, the conservative denies reality.
We mock them for it, but that is such low-hanging fruit. Or the
less they deny, the more they downplay the crisis, insist on further
research before transitioning energy sources, or caution against
regulation, all of which confess indifference. On the other hand,
the liberal who cries "believe the science" seeks to escape guilt,
vehemently lashing out against what needs to be done. If we need
to decarbonize the economy and dislodge capitalism, the liberal
says we must be more pragmatic! If the only candidate to propose
a marginally decent climate plan is a socialist, the liberal suddenly
remembers climate change is not so important. The moralizer
confesses: this is something that I should prefer to repress.

Moralizing is obnoxious but also dangerous. The symptoms
proliferate. This isn't a game. Don't you see what's on the horizon?

It is the late winter of 2020. Greece announces a plan to erect nets
in the Aegean Sea. There are too many Syrians. They are fleeing a
civil war that erupted as displaced people became concentrated in
hot cities during a drought. The nets will rise a meter out of the

---

10   Throughout the book, when I use the terms "the Changes" and
"the Dithering," I take inspiration from Kim Stanley Robinson, especially
his novel *2312*.

water to either catch or drown families. The crisis owes itself to a warmer climate. Nets don't discriminate.

It is 1978. Exxon scientists discuss the greenhouse effect and predict the collapse of the Western Antarctic Ice Sheet. Internal reports soon speak of widespread disaster, including a multimeter sea level rise and the flooding of Florida and Washington, DC. An alerted public might force them to strand assets and shutter operations. Exxon launches a misinformation campaign nearly a decade before the public learns the term "global warming."

It is one month before an election in 2018. The worst drought in decades sends a migrant caravan from the Northern Triangle north toward the southern United States border. They wish to claim asylum. Conservatives accuse the migrants of harboring Islamic State terrorists.

It is anytime in the so-called War on Terror. Drones strike school buses and wedding parties, and occasionally they hit resisting militants. Regions hit lie on the aridity line, the liminal zone—where barely enough rain falls for crops—that winds across the African Sahel through the Middle East. Plot the advance of desertification to find where drones might soon lurk. The Global North nervously whispers of approaching water wars, as if we are not already engaged.

It is 1989, when an American electric company schemes to justify a coal plant with a new tool. They plant millions of non-native trees in Guatemala and threaten locals with prison if they are cut down for firewood. Almost nothing is captured. The first "carbon offset."

It is 2004, when an oil company rebrands itself as green and launches one of the most successful advertising campaigns ever. BP teaches us to measure our iniquities by a "carbon footprint."

It is the now and the near future. Bangladesh's Kutupalong camp, the largest refugee camp in the world, is critically full of Rohingya driven out of Myanmar. The location is vulnerable to cyclones and floods. The government relocates a substantial portion of the population from the current camp in Cox's Bazar to an island called Bhasan Char. The low island is not twenty years old, formed from silt from the Meghna River. Already at high tide in a storm the island could be submerged—even before the seas rise, that is. Authorities will move hundreds of thousands of refugees to an island that will be drowned.

It is all happening so, so fast. The UN begs states to consider climate asylum cases, but there's no legal category for "climate refugee." A billionaire tests equipment to unilaterally fill the sky with aerosols to deflect sunlight, a step toward geoengineering that is, to say the least, controversial and dangerous. Murders of environmental activists reach new heights. States create laws to label activists as terrorists and suppress them with private mercenaries. Over a few months before the pandemic, liberal voters on both sides of the Atlantic reject leaders campaigning on the most ambitious climate legislation ever proposed. The Environmental Protection Agency (EPA) gives its climate leadership award to Raytheon. The Paris Agreement pretends there's hope for 1.5°C via nonbinding pledges that will push us over 3°C.

Indeed, the symptoms proliferate.

My thesis isn't complex: climate denial should be a flexible term designating a broad range of activity. I am not cleverly twisting a new definition. We find it intuitive to say teenage recklessness expresses a belief in invincibility (denial of mortality). Troubling symptoms are ignored or explained away by a patient in denial about what a physician might diagnose. Expensive purchases or self-sabotage during a midlife crisis are denials of fading time or unsatisfied desire. We commonly speak of denial as behavioral, but for climate change we reflexively confine ourselves to a narrow meaning of denial focused on conscious belief. Denial is not an incorrect thought to be transcended or an age to be passed. Dependence on hydrocarbons generates reactions or symptoms like an atavistic return of the repressed. More belief in correct ideas won't fix this. Denial is far more ramified and is expressed through material relations like pseudo-solutions and violence.

Stories and case studies in this book are conversions of denial—that is, behaviors consistent with what we might expect when people who don't believe in or care about climate change are suddenly confronted with its effects. How do imaginary antifa arsonists become a more intuitive explanatory heuristic than warmer conditions in the Pacific Northwest? Denial is best theorized as a tendency to negate threats in accord with material relations, generating symptomatic activity in order to justify, curate, and maintain regimes of power and material relations. I'm interpreting denial as would a psychoanalyst, where blockages, delay tactics, acting out, and even budgetary concerns are read as resistance. I hope to develop scientifically precise analyses within environmental humanities. The philosophers have interpreted the ecological crisis in various ways;

the point is to change it. Drawing on climatology, social sciences, and the humanities, this book builds a hybrid theory of denial for the left with a focus on languages, fantasies, and behaviors of corporations and governments.

We must learn the vicissitudes of denial, for we are running out of time. Human experience is overdetermined or layered with too many desires, incentives, confusions, and goals, and we need methods for defamiliarizing problems too casually ascribed to legible logics, sciences, or market incentives. In addition to the pleasure principle that directs life to enjoy and a reality principle that checks surplus pleasure-seeking (but in such a way as to forgo pleasure now to maximize pleasure later), Freud proposed an excessive drive operating beyond the pleasure principle: the death drive. Sadistic or masochistic activity can't always be read rationally, because we exhibit excessive libidinal drives and unconsciously encode our social systems and modes of production with the same excessive drives. What we are watching play out are death drives desirous of too many things at once.

My interest in climate change began modestly in college years in a conservative culture during the George W. Bush administration. "If God is going to destroy the world soon," I remember one minister saying, "why should we take care of it, even if we believed the scientists?" Or remember when Reagan's secretary of the interior, the late James G. Watt, was asked in a congressional hearing if he would defend the land for future generations and he famously replied, "I do not know how many future generations we can count on before the Lord returns."[11] Truth is impotent against shrewd theological desire, but both expressed a set of priors shared by half of American Christians at the time.

We approach climate change like Protestants, as if salvation depends primarily on whether an individual has the correct beliefs. My curiosity around climate change and apocalypticism persisted throughout my doctoral work in philosophy and religious studies. Anyone in these disciplines can tell you consciously expressed belief is mostly irrelevant to group behavior except insofar as beliefs are housed within ideological apparatuses holding levers of power. For example, white evangelical apocalypticism (literal future denial)

---

11 Robert D. McFadden, "James G. Watt, Polarizing Interior Secretary Under Reagan, Dies at 85," *The New York Times*, June 8, 2023, www.ny times.com.

resonates with the interests of capital (vague denial with a focus on short-term profit), and they form reactionary alliances in the Republican Party. As I documented in my book *Against: What Does the White Evangelical Want?*, around two-thirds of American white evangelicals, as well as somewhere between a fifth and a third of all Americans, say they don't believe there will be a twenty-second century.[12]

Apocalyptic nuts play an outsized role in the GOP and augment ecocidal policy, but it isn't clear a sudden change of heart could cripple either party's commitment to lease federal lands for extraction. Having exhausted this line of research, I turned to more clandestine and material expressions of denial.

The goal of this book is not to provide a full history of climate denial. For that, read Erik Conway and Naomi Oreskes's *Merchants of Doubt* and Kate Aronoff's *Overheated*. I won't engage heavily with the right's denial and its racist manifestations, because I consider the seminal work in this area to be Andreas Malm and the Zetkin Collective's *White Skin, Black Fuel*. Finally, I mostly ignore "doomers," those who overhype the dangers, and my close adherence to mainstream climate science should be read as a corrective to pessimistic impulses sometimes found on the left. I feel that pessimism, but there are many encouraging signs in the literature. For a variety of reasons, not least of which is the colossal role of my nation in contributing a fifth of historic emissions and raising surface temperature nearly a quarter degree as well as its ongoing derailment of mitigation efforts, I have drawn heavily on examples of denial in the United States. My goal is to map out recurrent patterns of denial expressed as government policy, corporate greenwashing and military or border violence.

Louis Althusser drew two conclusions about ideology. First, ideology represents an imaginary relationship to real conditions, aiding in recognition and misrecognition (denial does the latter). Secondly, ideology has a material existence. Ideology is housed within ideological state apparatuses (ISAs), assemblages allowing people to act out and reproduce an ideology. Althusser listed a few ISAs: the church or other religious institutions, public and private schools, the family, the legal system (both repressive and

---

12  See chapter 1 of Tad DeLay, *Against: What Does the White Evangelical Want?* (Eugene: Cascade, 2019).

ideological), political parties, unions, media, literature, arts, sports, and so on.

Ideology links a political and legal superstructure arising out of material relations of production that form the base of society. Denial is a heavy burden. You must do your part because our mode of production requires exploitation of labor and natural resources.

A simple taxonomy can clarify battle lines. Malm and the Zetkin Collective described Big Oil, front groups, and conservative think tanks in the seventies through nineties as an ISA coalescing to stall regulation.[13] Our first position is the *capitalist denialist* ISA.

Apparently in opposition but often in sync, a second position we'll call *capitalist climate governance* acknowledges the reality and the threat. But instead of a rapid phaseout of fossil fuels, it sets distant, nonbinding targets to protect profits. It neutralizes radical climate movements by co-optation or force, imposes no limit on accumulation, and creates new opportunities for profit in greenwashing. It suggests you look up your carbon footprint or purchase an offset. It abuses the IPCC's thorough assessments inasmuch as, per founding chair Sir John Houghton, the organization's remit is the tailpipe not the wellhead. "Talking about the source of fossil fuels would have moved us from the science area into the policy arena," Houghton said. "Because of the pressure we were under, we needed to be squeaky-clean, maybe too clean, but we needed it to be that way."[14] At best, capitalist climate governance delivers a limited legal and market architecture for transition. At worst, it's a haven for bad actors to greenwash accumulation.

In addition to the two owning class positions, there are two working-class positions. One is a proletarian *reactionary denial*, the position closest to the normal use of the term "climate denier." Today's reactionaries are increasingly ethno-nationalist and, rather than acknowledge the fires they see with their own eyes, will instead dial up the violence.

Since they are both conservative, the bourgeois capitalist denialist ISA and the proletarian reactionaries ally on the right. For example, when the Koch brothers funded the Tea Party movement

---

13 Andreas Malm and the Zetkin Collective, *White Skin, Black Fuel: On the Danger of Fossil Fascism* (London: Verso, 2021), 18.

14 John Houghton, as quoted in George Marshall, *Don't Even Think About It: Why Our Brains Are Wired to Ignore Climate Change* (New York: Bloomsbury, 2015), 171.

after the election of Obama, they took advantage of anti-Black racism to astroturf an apparatus conducive to deregulation. In what Ernesto Laclau called a "chain of equivalences," deregulation stitched itself to ethno-nationalism.[15] As conservatives mainstream QAnon-adjacent slander by casting opponents as groomers or part of a murky pedophilic cabal, obsessive thoughts about children and sex link ethno-nationalists to deregulatory agendas.

In other words, there are two alliances we face: a bourgeois alliance (capitalist denialist ISA and capitalist climate governance) and a conservative alliance (capitalist denialist ISA and reactionary ethno-nationalists).

What is needed is a fourth position, a *revolutionary* one, but of what use is a revolutionary position without the dialectical conditions for revolution? The characteristics are hazy, but let us say the fourth position detests the capitalist death drive and opposes the ethno-nationalist's solution. Bertolt Brecht said it best about his most minimal demand: whoever doesn't hope for the end of capitalism at this late date is an enemy of humanity "who does not think or who thinks only about himself."[16] The *telos* of the fourth position is a post-capitalist ecological socialism, but at the moment, climate activists display a wide ideological diversity. For now, this fourth position, mostly silent in this book and unsure what is to be done, is broadly characterized by anti-capitalist and egalitarian commitments. Without allies we must raise consciousness, including among reactionaries, and for the sake of those who will be harmed in the short to medium term, we must wring whatever concessions we can from capitalist climate governance.

My taxonomy can be shown as a semiotic square. Novelists use these for casting characters. Likewise, I use these positions not as an exhaustive list of four options but as a framework for mixed agendas. On the top there are two opposed terms ($S_1$ and $S_2$) and on the bottom two contradictions (not-$S_1$ and not-$S_2$). A semiotic square of denial advances a heuristic or shorthand to cast different roles.

---

15   See Ernesto Laclau, *On Populist Reason* (London: Verso, 2005).

16   Bertolt Brecht, "Communism Is the Middle Way," in *The Collected Poems of Bertolt Brecht*, trans. David Constantine (New York: Norton, 2018), 432.

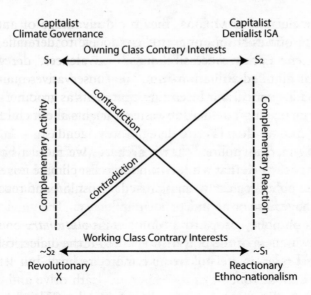

Figure 0.2. Semiotic Square of Denial

Capitalism isn't a roadblock for mitigation. This misrepresents its relationship to the Changes. No, capitalism is not a roadblock. Capitalism is the generator. The tendency of profit to follow the path of highest return, the pressure to lure investors with ever-greater profits quarter after quarter, the praise for unlimited expansion in a world of finite resources, labor exploitation and destruction of nature, the allure of cheap energy no matter the cost, and most of all the ability to discount true ecological and social costs that don't show up yet in the price of exchange-values —in such a deranged machine, climate change is a conversion disorder of capitalism.

Capitalism is an algorithm, or a set of directions or rules for projecting inputs upon a range of outputs. "Accumulate, accumulate, accumulate! That is Moses and the prophets," said Marx.[17] The law of accumulation is an independent social force, not a physical law, but nevertheless a vibrantly real phenomenon that requires neither villains nor plans to compel activity. Capitalism is a system defined by accumulation of surplus, an accumulation accomplished by exploiting workers and nature. The parallel exploitations generate class inequality and metabolic rifts. The capitalist reaps where she doesn't sow, as it were, twice. This crude logic of capital, which

17 Karl Marx, *Capital*, vol. I, trans. Ben Fowkes (London: Penguin Books, 1990), 742.

Marx formulated as M-C-M', means a drive to exploit labor and natural resources pushes us ever closer to a climate not seen since the Pliocene Epoch. Accessible fossil fuel reserves are worth a couple of hundred trillion dollars. Total fossil fuel resources are worth $2.4 quadrillion. Incentives point in the wrong direction. Climate change isn't a "market failure." It's a market success.

What can we do? There's that pronoun again. In a landmark study on American politics, the researchers Martin Gilens and Benjamin Page showed that while elite bourgeois and business interests determine policy regularly, "mass-based interest groups and average citizens have little or no independent influence." The impact of our opinions on policy drops to a "non-significant, near-zero level."[18] Public awareness of the polycrisis soars to unprecedented heights. Fossil fuel reserves are still worth hundreds of trillions of dollars. This isn't about you.

The algorithm takes as an article of faith what Theodor Adorno identified as the message of any horoscope column: "Everything will be fine."

There has never been such a monumentally difficult task, never such an existential threat, never such an interdisciplinary problem as the climate crisis. One knows the physics, another the sociology, yet another the carbon budgets. Leading experts occasionally gesture at an interdisciplinarity they not-so-secretly disdain. You see this in their demands to "leave science to the scientists," often in those exact words, even while proposing facile political solutions demonstrating a shallow reading of political philosophy or history. Marx warned of this problem:

> The weaknesses of the abstract materialism of natural science, a materialism which excludes the historical process, are immediately evident from the abstract and ideological conceptions expressed by its spokesmen whenever they venture beyond the bounds of their own speciality.[19]

We should read deeply where our expertise is slim. Nobody commands a god's-eye view of the polycrisis.

---

18    Martin Gilens and Benjamin I. Page, "Testing Theories of American Politics: Elites, Interest Groups, and Average Citizens," *Perspectives on Politics* 12, no. 3 (September 2014): 564–81.

19    Marx, *Capital*, vol I, 494.

"The class struggle," Walter Benjamin observed, "is a fight for the crude and material things without which no refined and spiritual things could exist." It is precisely the barest of all crude and material things that might be stripped from us and barred permanently. We will lose this world. As awareness rises, so, too, do emissions continue to rise, and pseudo-solutions waste precious time. We are discovering again the truth of the old Gramsci quote everyone butchers: "The crisis consists precisely in the fact that the old is dying and the new cannot be born; in this interregnum a great variety of morbid symptoms appear."

Will our descendants look back on the long twenty-first century as the Great Transition or the Dithering? Current production rates predict that we will burn through all oil and natural gas reserves around the time millennials become centenarians, with coal holding out a bit longer. The timeline will change as we quicken or slow the burn, or if we learn to extract resources we can't currently reach. Sure, the timeline is hazy, but the conclusion isn't. We will probably burn through most of our fossil fuel reserves. Capitalism will incentivize renewables inasmuch as they are profitable, not because the legal and political superstructure meets its unserious goals to stop the progress of this storm. In the meantime, the vulnerable suffer. Who has agency to switch algorithms? By the end of this book, you will need to decide whether even the revolutionary's desire for agency is denial.

Pathways to civilizational collapse are fundamentally unpredictable, nor can we divine social tipping points. Today's inchoate or incipient fascisms, as Richard Seymour rightly qualified this contested term, serve up a hint.[20] What will happen when hundreds of millions move, when economies plummet? How long until North America and Europe police their borders with armed drones, as Israel already does at the edge of Gaza? If the choice is socialism or barbarism, a fossil fascism will lash out nostalgically for a time before the seas swallowed the coasts while sneering at fleeing families: You're not from around here, are you?

"What was it like before the Changes?"

It is the early twenty-second century. The old man doesn't know how to answer the kid's question. She knows too much. Her

---

20 Richard Seymour, "Inchoate Fascism," Patreon, November 13, 2020, patreon.com.

textbooks recall sand covering ground like this everywhere, on every coast of every sea in the world.

When he was her age, he read up on the extinction of the rhinoceros. Lower Manhattan's carbon-based traffic before the seawall ruptured. Los Angeles before the fires. Southern Europe before the nets and razor wire. The Sahel and Northern Triangle before the aborted exodus. Oceans before the great dying. Bangladesh before the drowning. Her schoolbooks footnoted the last eagles or the first border sentry drones, sure, but the old man actually remembers events she reads like legends. Temperatures push ever closer to 4°C.

They sit on a beach in northern Michigan. He is my grandchild or yours, keeping watch over his family as best he can. The unnaturally red sun sets over the lake. In the winter, the waves still occasionally freeze like an eerie reminder of a world resisting control. Cold snaps and polar vortexes brutally compete against mass casualty heat waves, now a casual fact of life along with Category 6 hurricanes. She asks whether it was really true about the sand and smiles skeptically as he replies, "Before the Pulse, yes, there were undisturbed beaches like this everywhere."

After the greater ice sheets collapse in the West Antarctic at the end of the twenty-first century, the world scrambles. It was too much, on top of Greenland's melting and ice-free summers at the North Pole. A classic theological problem: What kind of gods foresee sin yet refuse to act? Something about free will. No, not satisfying. The first pulse raises the sea level nearly two meters over three decades.

"Even on the coasts there was sand?" she asks. Not accusative, her tone. She doesn't know who to indict.

For years, the "managed retreat" of beach towns went in fits and starts. After the Pulse, though? Every port and coastal city in the world is inundated. The tide pushes sand up into abandoned streets of desolate cities, swamping estuaries and choking grass and trees in salt. New high rises built on higher ground tower over shoreline ruins.

Food is already prohibitively expensive as crop yields on once-dependable farmland drop by half, requiring expansion to less fertile land in a world struggling and failing miserably to feed ten billion stomachs. Rations could be printed, but subscription fees keep the unemployed starving.

Capital trudges along and charts new ways to reap profit from fires and wars even as economies plummet into never-ending crises.

Nationalists respond with growing demands for trade tariffs. Louder still are their roars for blood. Economic growth stalls out below 1 percent, and hopes of lessening inequality disappear. The start of the permanent depression, the normal and the frightening kind.

"I wish you could have seen it," he replies.

The oligarchs close ranks and retreat to bunkers or walled cities. New fiefdoms. Amusement parks for their rotting souls. Sometimes, guards turn on the lords.

Communes spring up near fresh water, which is at least somewhat insulated from the melting. Some thrive. Most are suppressed by legal tricks of eminent domain or privatized police. The last undisturbed beaches in North America, along with the fresh water, draw waves of migrants north even as temperatures scorch the Midwest.

What else is there to say? Too many are displaced for reasons environmental or economic, to say nothing of the balkanization. None but the luckiest live near family. Aside from birds, most children haven't even seen a wild animal. The skies are milky white, never blue anymore, because of the stratospheric aerosols released to dim the sun.

See, the Great Lakes are clustered aftermath of glacial retreat ten thousand years back. Glaciers covered much of North America at one point. When they melted away, they left rich soil deposits. Further back, shallow waters covered the American South during the Cretaceous Period. Sea creatures died and their calcium carbonate exoskeletons accumulated. As carbon dioxide dropped and polar ice caps formed, the seas withdrew and left fertile soil. Millions of years later, that fertile soil of the Black Belt would draw slaveholders to force the planting of cotton, tobacco, sugar, you name it. This set off a chain reaction of voracious capitalism and reactionary politics that, by the time climate change was discovered, stood poised and ready to murder every ecosystem rich with resources, which was all of them. Very lucrative.

"Could there be beaches again, like on the ocean?" she asks innocently. Her grandfather weighs whether to wax poetic about the earth's ability to heal from ruin, the carbon drawdown cycles operating on timelines of thousands and millions of years. What a convoluted, cruel hope. They stole so much.

"No."

# PART I

*Denial*

# 1

# How to Deny a Catastrophe

*The only responsible option is to deny oneself the ideological misuse of one's own existence, and as for the rest, to behave in private as modestly, inconspicuously and unpretentiously as required, not for reasons of good upbringing, but because of the shame that when one is in hell, there is still air to breathe.*

—Theodor Adorno

*Truth is not in desire's nature.*

—Jacques Lacan

The Sevenfold Litany will set out at dawn. Winding through Roman streets the procession will reach the Basilica of Saint Mary Major. They will plead for the virgin mother's intercession. A pestilence afflicts the city in spring 590 CE.

The pope is lost to the disease. Into the void steps a deacon, known to history as Pope Gregory the Great. He warns his flock their iniquities are the cause. They are "struck by the sword of God's anger, smitten down by sudden death." If unfaithfulness has incurred judgment, penitence should rectify.

He orchestrates marches across the city. Clergy and priests begin at one church, monks from another, and nuns from yet another. Children would approach from here, laymen from there, and finally widows and married women begin at their separate points. All converge upon the basilica.

In the midst of a contagious pestilence, bodies corral together, bumping into one another, breathing on each other. The infected seek relief too, of course, so they march with the others in false hope. History records scores of penitents fallen just in those hours marching.

Is it worth the casualties? In the estimation of the Sevenfold Litany, yes, because God stays his judgment. Gregory's gamble pays off, not only growing his fortunes when he is soon elevated to pontifex, but also lifting the stature of the bishops of Rome in perpetuity as popes. A useless gesture, almost certainly what we would now call a super-spreader event, is credited with saving Rome. Humans hallucinate salvation in nonsense.

## Analogies

At the start of the pandemic, I noticed analogies drawn between Covid-19 and either plague or climate change. Such talk disappeared months in, which is interesting for different reasons, but the heuristics in these analogies give us a useful shorthand for denial.

Covid-plague analogies portrayed mask or vaccine skeptics as morons incapable of risk assessment, unlike pre-modern yet wiser Europeans during the Black Death. If illiterate medieval peasants took prophylactic measures, why couldn't Covid-19 deniers?

Covid-climate analogies compared drastic measures to limit suffering (lockdowns, direct cash payments, rapid vaccine development, social distancing, virtual education) and far more costly investments needed for the climate crisis. If we were willing to invest and discipline ourselves, didn't graver problems deserve more sacrifice?

Responses to the pandemic turned into volleys in a culture war. Wearing a mask was a sign of allegiance. "Believe the science" pulled double duty for pandemic and climate concerns, but none of us read the science (who has the time?) and, instead, cobbled together fantasy-delusions from social media and vibes. Some science believers pledged to wear masks forever to avoid being mistaken for a conservative, while conservatives coughed at service workers as a fun death-threat game. All individual reaction and moral puzzles.

Out of sight, the capital algorithm ordered us back to work. It slashed unemployment payments, authorized evictions, and framed sickness as moral failure. If people choose poorly, so neoliberal judgment proceeded, they have themselves to blame! Cruel, but we can use short-sighted Covid-plague and Covid-climate analogies paired with classic responses to plague as a heuristic for studying denial.

## The Plague of Justinian

By the end of the cold winter of 542 CE, the bacterium *Yersinia pestis* arrived in Constantinople. "At first only a few people died above the usual death rate but then the mortality rose higher until the toll in deaths reached five thousand a day," wrote the historian Procopius. "And after that it reached ten thousand, and then even more."[1] Markets emptied of food and commerce ceased. Those venturing from home wore identification tags in case they fell dead.

In a city of a half million people, ancient sources estimated 230,000 to 300,000 perished. Even by antiquity's inflated statistics, the toll was staggering. *Y. pestis* later drew more infamy as the Black Death in the fourteenth century, but, in the sixth, the plague first advanced into Europe. Emperor Justinian caught the disease and survived. The wealthy expect they're insulated from calamity, and not without reason. Justinian delegated relief efforts to court officials and treated a plague known to history by his name as a nuisance.

The outbreak was a climate event. The world was at its coldest in two millennia. For the last hundred centuries of the Holocene, Earth's inhabitants enjoyed an unusually stable interglacial period. Shocks were rare, dangerous. From 200 BCE to 150 CE Europe enjoyed the Roman Climate Optimum. Borders shifted with victories owed to warm and wet conditions suitable for an agrarian society supporting trade and force. After the Optimum came the Late Antique Little Ice Age, when temperatures fell 2.5°C. In the eastern empire, Constantinople rode out the cold thanks to grain storehouses. Humans and rats feasted on the rations.

By 536 CE, they saw the changes. John of Ephesus reported, "The sun darkened and stayed covered with darkness a year and a half, that is eighteen months. Although rays were visible around it for two or three hours (a day), they were as if diseased."[2] Dimming from a brief eclipse was frequently seen as a harbinger of evil, but this was no brief eclipse. Imagine the distress of a months-long darkening. "During the whole year the sun gave forth its light without brightness, like the moon, and it seemed extremely like

---

1   Procopius as quoted in Kyle Harper, *The Fate of Rome* (Princeton: Princeton University Press, 2017), 226. I also wish to credit Harper with examining the Plague of Justinian as a climate event.

2   John of Ephesus as quoted in Harper, *The Fate of Rome*, 251.

the sun in eclipse," Procopius wrote. "From the time when this thing happened men were free neither from war nor pestilence nor any other thing that brings death."[3] The panic doubled down on destruction.

Christians saw signs of God's judgment in the dimmed sun. Praetorian Prefect Cassiodorus blamed Justinian's Roman renewal, since catastrophe follows "when kings change their established customs."[4] Poor individual choices and sin were culprits.

The truth was a massive volcano erupted in 536 CE in the northern hemisphere. It kicked up so much sulfur dioxide that the sun's light was blotted the world over. Then, in 539 or 540 CE, another eruption from a tropical volcano plunged Earth deeper into cold.

The exact relation of the Little Antique Ice Age to plague remains unclear, but proximity hints at causal relationships. *Y. pestis* was carried by a flea species that feeds on the black rat. As rat populations dwindled and fleas switched to human blood, other flea species primarily feeding on humans carried plague as well. Cold rats scurried into warm human homes. The flea, the rat, the bacterium, and the family sheltered together in human homes.

By the time it reached Constantinople, plague had completed a journey hundreds of years in the making. *Y. pestis* diverged from *Y. tuberculosis* tens of thousands of years ago, and modern *Y. pestis* evolved at least a thousand years before the outbreak in 542. It had no efficient method of travel until the trade routes of the Roman Empire and the Silk Road. Plague arrived in the Suez just as Egypt, the empire's breadbasket, stepped up grain shipments.

Crops failed and threatened disorder. Alexandrian storehouses moved hundreds of millions of liters of grain to Constantinople. Rats and fleas sailed along. The storehouses opened.

The plague afflicted untold casualties across several continents. Regular outbreaks of *Y. pestis* persisted for hundreds of years before quieting until the fourteenth century.

Ironically, it was preparation for cold that generated the vulnerability. Exposed to ecological flukes and flaws in resource distribution, they were so fragile, so unwittingly vulnerable to a couple of degrees.

---

3  Procopius as quoted in ibid.
4  Cassiodorus as quoted in ibid., 252.

## Symptoms

Conflicts in the psyche often manifest elsewhere in the body. This is no longer controversial. One need not be a psychoanalyst to see it. So-called cognitive or evidence-based psychologies say the same thing. Modern medicine agrees, at times calling it conversion disorder or perhaps functional neurological disorder. But psychoanalysts discovered it as early as the talking cure.

The conversion symptom is an old idea worth a hearing again. Sigmund Freud and his colleague Josef Breuer studied how the latter's patient Anna O. suffered physical symptoms from mental distress. She dreamed a snake attacked her father, and ever after, the sight of a snake caused her arm to stiffen. She sporadically lost her ability to speak her native tongue, yet she retained use of another. These were conversion symptoms.

The conversion symptom is a return of the repressed, from psyche to the body. The psychoanalyst Bruce Fink listed "minor aches and pains, tightness in the chest, a tingling sensation, a burning sensation, and dizziness to migraines, paralysis, blindness, muteness, and deafness."[5] Today those resistant to psychoanalytic thought still complain of stress-related irritable bowel disorder.

Of what am I trying to warn? Capitalism's carbon dependency creates denial in the form of pseudo-solutions, gimmicks, false promises, and quotidian violence much like repression leads to a symptomatic return of the repressed. The formula Marx identified as M-C-M′, where money transforms into a commodity, which turns back into more money, means capitalism accumulates surplus from labor, which transforms nature. Wealth comes from labor and the land. The law of accumulation points to expansion, not long-term stability or even short-term use value. It plans in financial quarters or, at most, in twenty-year bonds. It doesn't plan for human flourishing. In a deranged machine, climate change is like a conversion disorder of capitalism and denial is its symptom.

A symptom indicates an underlying conflict or repression, for which Freud used a timeless case study: the *fort-da* game.[6] According to Freud, at the age of one and a half, his grandson could say

---

5 Bruce Fink, *A Clinical Introduction to Lacanian Psychoanalysis* (Cambridge: Harvard University Press, 1999), 114.

6 Sigmund Freud, "Beyond the Pleasure Principle," in *The Standard Edition*, vol. XVIII, trans. James Strachey (London: Hogarth, 1955), 14–17.

only a few words, slept well through the night, and bonded strongly with his mother, though he didn't protest when she left the room. He often tossed toys to a corner and shouted "o-o-o-o," trying to say "*fort*" (gone). Holding a string attached to a wooden reel, the boy threw the toy from his crib shouting "o-o-o-o" before yanking the string to make the reel reappear while shouting "*da*" (there). *Fort* and *da*, *fort* and *da*. The disappearance and reappearance were both parts of the game, but the greater pleasure came from the second act. On Freud's interpretation, in his passive helplessness the child felt his mother's departure as stressful, so he compensated with an active game.

The psychoanalyst Jacques Lacan added the boy clearly wasn't replacing the mother with the *fort-da* game. He knew she was gone yet coped with a game. The game was a representative in the place of a (missing) representation.[7]

This is how repression works: when two ideas or drives conflict, rather than one being preferred over the other, one idea becomes unconscious and generates symptoms while the other operates consciously. The boy wanted the mother, yet he also wanted security in her absence. Unable to satisfy both, her absence was made unconscious and he generated a game, an innocent symptom. According to Lacan, "repression and the return of the repressed are just the two sides of the same coin."[8]

Surely guilt for our carbon and easy lifestyles built on the suffering of others is among the feelings we repress. Symptomatic returns of the repressed exacerbate injustice: foolishness pseudo-activity, moral distancing, and violence. Once repression starts, ever more frantic activity is marshaled against the unconscious becoming conscious. Like Freud said, "Repression is not an event that occurs once but that it requires a permanent expenditure."[9]

---

7  Jacques Lacan, *The Four Fundamental Concepts of Psychoanalysis: The Seminar of Jacques Lacan, Book XI*, ed. Jacques-Alain Miller, trans. Alan Sheridan, (New York: Norton, 1998), 60–3.

8  Jacques Lacan, *The Psychoses: The Seminar of Jacques Lacan, Book III*, ed. Jacques-Alain Miller, trans. Russell Grigg, (New York: Norton, 1997), 12.

9  Sigmund Freud and James Strachey, "Inhibitions, Symptoms and Anxiety," in *The Standard Edition*, vol. XX, trans. James Strachey (London: Hogarth, 1959), 157.

## The Big Death

*Yersinia pestis* reemerged in 1347. Inside five years, twenty-five million of Europe's seventy-five million people perished, along with as many as two hundred million worldwide.[10] Not until three centuries later did the historian Johannes Isaacus Pontanus ascribe to it the name Black Death. At the time, people used an almost childlike name. The historian John Kelly listed names in European tongues. "*La moria grandissima, la mortalega grande, très grande mortalité, grosze Pestilentz, peligro grande*, and *huge morta-lyte* . . . 'Great Mortality,' or, more colloquially, the 'Big Death.'"[11]

It spread too fast. Nobody knew why. Not long after the plague, the English monk Henry Knighton explained it as God's judgment for poor behavior of young women. He noticed outbreaks near tournament locations and deduced that women dressing up in men's clothes and following the games, "the showiest and most beautiful, though not most virtuous, women of the realm," must trigger divine wrath. He whined about the gold and silver adorning their immodest hips as he snickered at how "God, present in these things, as in everything, supplied a marvelous remedy."[12]

Educated persons knew better. They blamed it on *miasma*, or bad air, drawing on the ancient medical wisdom of the Roman doctor Galen. The Florentine polymath Giovanni Villani pointed to the stars, insisting plague was "foretold by the masters in astrology last March."[13] Parisian medical faculty backed up such views. Their compendium traced the outbreak to the exact hour of "a major conjunction of three planets in Aquarius."[14] Many blamed Jews.

The black rat and its fleas evaded blame. So did climate. In the early fourteenth century, perhaps at the early edge of the Little Ice

---

10   John Kelly, *The Great Mortality: An Intimate History of the Black Death, the Most Devastating Plague of All Time* (New York: Harper Perennial, 2006), 11–12. Miscellaneous details on plague responses also from Joshua Mark, "Religious Responses to the Black Death," World History Encyclopedia, April 16, 2020, worldhistory.org. See also Joshua Mark, "Reactions to Plague in the Ancient and Medieval World," World History Encyclopedia, March 31, 2020, worldhistory.org.

11   Kelly, *The Great Mortality*, 23.

12   Henry Knighton as quoted in ibid., 18.

13   Giovanni Villani as quoted in ibid., 103.

14   *Compendium de epidemia per Collegium Facultatis Medicorum Parisius* as quoted in ibid., 169.

Age, unusually cold and wet summers meant floods, poor harvests, and skyrocketing food prices. By 1316, so many starved that in the streets of Antwerp one could hear men making the rounds yelling, "Bring out your dead!"[15] Contemporaries saw connections between malnutrition and susceptibility to disease. Environmental factors are less certain than zoological causes, but they were present in both the Plague of Justinian and the Black Death.

Kelly summarized four responses to the plague, drawing on Giovanni Boccaccio's *Decameron*, a work of fiction written after and inspired by Boccaccio's experience surviving the Big Death.[16] Notice in them familiar responses to Covid-19 and premonitions of how we might respond to the climate crisis.

First, many retreated from others and isolated themselves, understanding plague to be communicable and thus social interaction a mortal risk. Boccaccio described characters "of the opinion that a sore and abstemious mode of living considerably reduced the risk of infection," so instead they "formed themselves into groups and lived in isolation from everyone . . . consuming modest quantities of delicate food and precious wines and avoiding all excesses."[17] Villani said some figured infection spread by sight: "Early on men, women, and children saw that the disease could strike simply by touch, even sight, and could be recognized by the tell-tale traits of the swelling."[18]

Second, many ignored caution and satisfied carnal cravings. In addition to wine and parties, treating plague as, in Boccaccio's words, "one enormous joke," raucous crowds curated their speech to ignore the plague. "Conversation was restricted to subjects that were entertaining or pleasant," since their days were numbered. They "believed that drinking excessively, enjoying life, going about in song and celebration, satisfying in every way the appetites as best one could, laughing, and making light of all that happened was the best medicine for such a disease."[19] Cynicism and irony defended against despair.

Third, some took a middle course between isolation and carelessness and preferred caution or risk mitigation. Boccaccio reported

---

15    Ibid., 58–61.

16    Ibid., 105–11.

17    Boccaccio as quoted in ibid., 109.

18    Villani as quoted in Samuel Cohn, "Plague Violence and Abandonment from the Black Death to the Early Modern Period," *Annales de Démographie Historique* 134, no. 2 (2017): 39–61.

19    Mark, "Reactions to Plague in the Ancient and Medieval World."

people carrying flowers and herbs in their hands or spices shoved into their noses, "believing that such smells were wonderful means of purifying the brain, for all the air seemed infected with the stench of dead bodies, sickness, and medicines."[20] The bird-beaked plague doctor filtering respiration with scents stuffed in a mask is a later invention, but potions and scents were common mitigation technologies. At home, various aromatics warded off the bad air, while out in the world, people carried smelling apples. Some believed sex and baths opened the pores and exposed one to disease. Changes to diet, the consumption of wine, and the avoidance of wine were tried as well.[21] A popular remedy was to purchase a unicorn horn and grind it up into a potion to drink. Some closed windows facing the direction from which they believed the bad air blew. Others cut up a serpent as a symbol of the devil and rubbed its carcass on their skin to draw out the evil in the buboes. The most amusing innovation in risk mitigation came from the physician John Colle, who noticed "attendants who take care of latrines and those who serve in hospitals and other malodorous places are nearly all to be considered immune."[22] Imagine the lords sniffing the latrines.

The final response was desertion. Those with means comfortably fled infected cities for the countryside while impoverished peasants moved to cities in hopes of work or better medical care should they fall ill. Families abandoned one another. Bleak records repeat sad scenes of parents abandoning infected children and vice versa. "This scourge had implanted so great a terror in the hearts of men and women that brothers abandoned brothers, uncles their nephews, sisters their brothers, and in many cases wives deserted their husbands," Boccaccio wrote. "But even worse, and almost incredible, was the fact that fathers and mothers refused to nurse and assist their own children, as though they did not belong to them."[23]

Locales variously treated plague as a logistical problem, a moral puzzle, or a non-issue. Committees across the Italian peninsula pioneered what we would today call public health. They screened incoming ships, shuttered nonessential businesses, and adjusted

20   Ibid.
21   Kelly, *The Great Mortality*, 172–3.
22   John Colle, as quoted in ibid., 173.
23   Boccaccio as quoted in Cohn, "Plague Violence and Abandonment," 43.

funeral rituals. Elsewhere, priests and public officials abandoned duties as civil society collapsed. One town banned gambling, collected money for the poor, and proclaimed a religious festival to win back God's favor. In still other places, the plague was practically ignored. Records of the Council of Seven, a governing body of Orvieto, fully ignored plague except for a single reference nearly a year into the outbreak once disease engulfed the town and six of the seven lay dead.[24]

The Church saw plague as either God's punishment or holy instruction, mirroring the miserable explanations of the Book of Job to the point of parody. Bishop Ralph of Shrewsbury said, "Almighty God uses thunder, lightning and other blows . . . to scourge the sons he wishes to redeem." The Heavenly Letter, a popular document circulating during the plague, castigated the floundering faithful, "Ye have not repented of your sins . . . Thus, I had thought to exterminate you and all living things from the earth; but for the sake of my Holy mother . . . I have granted a delay."[25] Some Muslim clerics taught that the plague was God's will or that succumbing was a type of martyrdom and, by some reports, viewed life and death during plague as the prerogative of God alone.

Moral masochism, an excessive enjoyment of self-criticism or humiliation, is one of the twenty-first century's clearest recapitulations of the fourteenth. Spectacular troupes of fifty to five hundred flagellants toured central Europe. The movement began nearly a century prior, but plague turned the spectacle of penitential self-harm into common entertainment. Into afflicted towns men marched as church bells rang beckoning all to witness shoeless and stripped troupes bruising themselves as they walked. As their ceremony proceeded, the men all at once collapsed into positions indicating their sin (adulterers lay prone, perjurers held several fingers over their heads, and so on). A leader whipped the men until they stood in unison and beat themselves bloody while singing a hymn proclaiming their salvation from hell. Wherever the flagellants went, so went persecution of the Jews.

In Europe, the Jews suffered worst. At Avignon, Christians wouldn't eat fish out of concern their protein was contaminated with plague from Genoese ships, and at the same time, the denizens

---

24  Kelly, *The Great Mortality*, 98.
25  Bishop Ralph of Shrewsbury and the Heavenly Letter as quoted in ibid., 222, 265.

burned their Jewish population, as they did in Strasbourg and other towns, as a type of prophylactic measure. Fearful of contaminated water, the town of Speyer killed its Jews but sent corpses downriver in barrels. Many were tortured and beat to death. Reportedly in Worms, sensing their imminent murder, some Jews locked their doors and burned down their houses while still inside.[26]

Pogroms in the past erupted at Easter with myths of blood libel. Christians accused Jews of capturing children and inflicting twisted rituals of torture. During the plague, there was no day or night to barbarism. The first pogrom of the plague occurred on April 13, 1348, during Holy Week, where in Toulon, dozens of Jews were murdered.

Over the next few months, Christians spread rumors of Jews poisoning wells, which ballooned into an international conspiracy for world domination. Pope Clement VI pushed back cautiously, arguing Jews were dying of plague just as quickly as anyone else, and he announced anyone attacking Jews should be excommunicated. At the same time, Holy Roman Emperor Charles IV canceled debts owed to Jews and goaded his subjects into murder. Economic expropriation ran parallel to physical violence. That's why violence was driven by elites rather than peasants.

The origin of the well-poisoning conspiracy theory appears to be as follows. In the Swiss town of Chillon, authorities arrested a Jewish surgeon who, according to his torturers, confessed to the existence of a plot to poison Christians, led by a figure called Rabbi Jacob. Out of this nascent delusion sprung a whole cast of supporting characters carrying out various tasks to wipe out Christians. The amount of poison required was said to be the size of an egg, or a nut, or two fists. It was made from lizards or spiders, sometimes from human hearts, or maybe communion wafers.[27] From Chillon, the rumors spread across Europe in months. Jews suffered a double threat of death by plague or fire. We still do not know how many perished.

Lessons are many. Among them is the enjoyment of pseudo-solutions and barbarism. Most importantly the lesson is this: people responded in denial when they didn't have meaningful agency within regimes of power under which they suffered.

---

26  Ibid., 26.
27  Ibid., 139.

## Negation of Reality and Negation of Moral Culpability

The ego constantly censors unpleasant thoughts from consciousness. In Freud's analysis of negation (*Verneinung*),[28] the patient offered up, "You ask who this person in the dream can be. It's *not* my mother," which Freud interpreted as, "So it *is* his mother." Amending the statement to its opposite was only the simplest formation of denial.

When the repressed toggles to conscious, if what surfaces feels unpleasant, the patient negates. In order: first, a repression; second, the repression begins to falter; and third, a need to negate whatever threatens to become conscious.

In another example, a neurotic approached Freud and said, "I've got a new obsessive idea, and it occurred to me at once that it might mean so and so. But no; that can't be true, or it couldn't have occurred to me." Here, we have the basic structure of denial: the repressed idea becoming conscious, then the idea's negation. Negation attests to the truth of the repudiated idea. Freud arrived at his formula: "Negation is a way of taking cognizance of what is repressed; indeed it is already a lifting of the repression, though not, of course, an acceptance of what is repressed." The emotive affect and the idea split and allow the affective or emotional component to remain repressed even as the patient consciously acknowledges the idea (or vice versa).

If we were constantly bumping up against repressed material without a way to flex or dynamically respond, our psyches would be prisons. Splitting affect from idea through negation, or peeling off what we feel from what we know, allows a bit of leeway. That's what a negative judgment does. To judge an act or thought negatively is akin to deciding, "This is something I should prefer to repress." Freud saw two categories of judgment that are crucial to our study of climate denial.

First, a judgment "asserts or disputes that a presentation has an existence in reality." All representations in the internal mind started as some sort of external real, but the ability to remember something that once happened means we have the ability to call forth an image or representation that is not, at least currently, real. Humans constantly call up memories to distort, test, or reject.

---

28  Sigmund Freud, "Negation," in *The Standard Edition*, vol. XIX, trans. James Strachey (London: Hogarth, 1961), 233–9.

Second, a judgment "affirms or disaffirms the possession by a thing of a particular attribute." Freud described this second rejection-judgment as moral. Originally, moral judgment might have been about practical usefulness or harm, such as "good" or "bad" food I wish to take into my body or spit out. The pleasure-seeking ego sought to "introject into itself everything that is good and to eject from itself everything that is bad." And initially, bad means external or alien.

Climate denial is negation and expresses one of these two negating judgments. One tests and negates reality itself, while the other negates and rejects moral culpability or alien ideas. Each signifies a failed repression.

Denial of moral culpability feels seductive, because you can't admit, as a capitalist subject, that there's little you as an individual can do, and neither can you imagine the end of capitalism. Liberals tend toward the rejection of moral culpability while conservatives usually reject reality, but the truth is we are now awash in so many options for negation it becomes difficult to separate denial from authentic mitigation. While reactionary reality denial still hammers old myths about deceptive scientists or sun cycles, moral denial asks for carbon offsets, hybrid cars, local purchases, recycling, and so on. That's not to say all behaviors to lower my guilt over environmental impact are foolish (lowering my carbon contribution is minimally beneficial), but authentic mitigation and false greenwashing develop in parallel as moral negation.

**Reality Negation**

In fall 2018, a fire sparked and destroyed more than 150,000 acres. On November 8 the Camp Fire devastated a town called Paradise in a matter of hours, too quick for many to hear the evacuation order. At least eighty-five people died. The California Department of Forestry and Fire Protection eventually determined the source to be electrical transmission lines operated by the Pacific Gas and Electricity Company. This source and the exacerbation of Californian fires by a warmer climate are not in dispute, but future US congressional representative Marjorie Taylor Greene declared a different theory.

Greene announced on Facebook her investigation into wildfires. "There are too many coincidences to ignore," she hinted. Democrats and Jewish bankers, she said, colluded to drive up the stock price of the electric company while clearing the countryside for high-speed trains. The rail costs so much, she lamented, "we could build three US southern border walls." So they were using satellite platforms to beam solar radiation into the forest. "If they are beaming the suns [sic] energy back to Earth, I'm sure they wouldn't ever miss a transmitter receiving station right??!!"

She pressed her case. "What would that look like anyway? A laser beam or light beam coming down to Earth, I guess. Could that cause a fire? Hmmm, I don't know. I hope not!" she asked of imaginary culprits she's outed. "Hmmm, I don't know." Once more blaming liberals and Jews for space lasers rather than climate change, she feigned humility, "But what do I know? I just like to read a lot."

Ideology represents an imaginary relationship to real conditions, aiding in recognition and misrecognition. This was Louis Althusser's first thesis on ideology. Denial or reality negation is misrecognition. When misrecognition is deranged, we tend to laugh it off rather than learn the contours of ideological commitment. Jewish space lasers were a more intuitive leap for a future congressperson than a flaw in the power grid in a hotter climate.

The problem with confronting denial head on is the same as with any symptom: derived symptoms or ideas are easily switched out. Greene eventually forgot about lasers and devoted energy to the QAnon conspiracy in which a righteous Donald Trump would purge a ring of liberal, satanic pedophiles. Correcting either nonsense would have launched the next delusion.

Without misrecognition, too much knowledge would pierce and dissolve the imaginary relationship. Slavoj Žižek's definition of ideology builds on Althusser's but, crucially, distances the subject from knowing too much.

> This is probably the fundamental dimension of "ideology": ideology is not simply a "false consciousness," an illusory representation of reality, it is rather this reality itself which is already to be conceived as "ideological"—*"ideological" is a social reality whose very existence implies the non-knowledge of its participants as to its essence*—that is, the social effectivity, the very reproduction of which implies that the individuals "do not know what they are doing." *"Ideological" is not the "false consciousness" of a (social) being but this being itself in so far as it is supported by "false consciousness."*[29]

29  Slavoj Žižek, *The Sublime Object of Ideology* (London: Verso, 2008), 15–16.

There's no escaping what Žižek called the social reality of non-knowledge, what Althusser called the imagined relationship to real conditions, or what psychoanalysts call fantasy.

## Moral Negation

Althusser's second thesis on ideology—ideology has a material existence—meant ideology is housed within apparatuses, allowing populations to act out and reproduce an ideology. Althusser rightly called the school an apparatus reproducing ideology. Schools exploit and repress while they teach and liberate. Schools implement dominant virtues, as does a house of worship, the family, the military, popular books, television, and the sports stadium. As Marx said, the ruling ideas are always those of the ruling class. In a capitalist society whatever constitutes a virtue, which is to say whatever is advantageous for ruling classes, is disciplined into working classes.

This brings us to a conjunction of Althusser and Freud. If ideological apparatuses reproduce certain virtues (a moral term), and if subjects fight against the repressed becoming conscious with moral negation, then we should expect to see people regularly straining with awkwardly alien moral logics to distance themselves from blame for the climate crisis. The temptation is greatest for proletarians trying to conform their consciousness to the thinking of the bourgeoisie or, to bring this to the subject at hand, to the solutions available according to capitalist climate governance. We should expect to find working classes and professional-managerial classes replicating whatever moral language is currently in vogue among businesses portraying themselves as responsible actors.

Before dawn on July 2, 2021, the ocean caught fire just west of Mexico's Yucatán Peninsula. Near a state-owned Pemex drilling platform, a circular eye of fire erupted at the water's surface. The waves themselves appeared molten orange and yellow. It was unreal, more like a gate to hell than an industrial accident. Video spread quickly over social media. I distinctly remember scrolling past the footage several times when mistaking it as an ad for some forgettable Hollywood catastrophe.

The fire began from a pipeline. Nobody was sure how long gas leaked before igniting. The company blamed a lightning storm and deflected with a statement about oil: "There was no oil spill and the immediate action taken to control the surface fire avoided environmental damage."

Two dynamics caught my interest, one, of course, being the spectacular images of an ocean on fire just as I was in the early days of writing a climate book. The other was the moralization around the incident. While plenty of environmental groups such as Greenpeace Mexico accused Pemex of "ecocide," a loaded but apt term, others grasped in the dark for a culprit. In one viral social media comment, a popular author confused the ecological crisis with the issue of gender inequality: "Men had their chance. This is what they did with it. It's time for women. Progressive women."

By weaponizing a vulgar idea of identity and oppression, ruling classes easily cast victims as perpetrators. When US vice president Kamala Harris told Guatemalan president Alejandro Giammattei to slow migration, she mixed up a word salad of liberal-friendly concerns. "There are also longstanding issues that are often called the 'root causes' of immigration," Harris explained to the leader of Guatemala, a nation the United States repeatedly crushed under its boot. "We are looking at the issue of poverty and the lack, therefore, of economic opportunities; the issue of extreme weather conditions and the lack of climate adaptation; as well as corruption and the lack of good governance; and violence against women, Indigenous people, LGBTQ people, and Afro-descendants." The vice president cast migrants as leaving in protest at homophobia, violence, and racism, to which, presumably unlike the United States, Guatemala was insufficiently responding.

The analysis of climate change requires we keep multiple immense problems in mind simultaneously: climate science, energy, racism, capital. Simple enough to analyze any two together, it's far more difficult to hold all four in mind at once. Liberals often default to mistaking climate change as an identity problem, but framing the problem as one of gender, race, age, or sexuality as such (rather than theorizing gender or race as crucial components of capitalist superstructure) occludes how liberal capitalist order depends on racism as a line for distributing resources and harm. But this is far too abstract for a vice president blaming the Guatemalan government's homophobia and racism to deflect from US oppression. These mistakes of attribution, which obscure real causes of bourgeois power within the capitalist algorithm, worsen violence against Indigenous communities.

Having established a framework for denial, we turn to the strange vicissitudes of pseudo-solutions. If keeping fossil fuels in the ground is off the table, who else can we blame? How can reality negation and moral negation help? Repression takes constant energy, and when repression fails, the negations lash out in a sprawling theater of accusations, misrecognitions, barbarism, and trivial grifts seeking purity.

# 2

# Footprints and Offsets

*The more your quest is located on the side of truth, the more you uphold the power of the impossibles . . . governing, educating, analyzing on occasion.*

—Jacques Lacan

It is a new millennium. British Petroleum rebrands, now using the lower case "bp" in its logo to stand for "beyond petroleum." Nice. Unlike ExxonMobil's strategy of science denial, BP positions itself as the first green oil company. Adopting a new logo in a green, yellow, and white sunburst, they pioneer a field variously called environmental advertising, greenwashing or green spin.

"What size is your carbon?" asks black text in the ad blitz. Fade to interviews on the streets of London. "Ah, the carbon footprint, eh," one respondent pauses, "that I don't know." Another reframes the question, "How much carbon I produce—is that it?" A final interviewee in Chicago rounds out the explanation. "You mean the effect that *my* living has on the Earth in terms of the products I consume?" The ad fades back to white field, black text. "We can all do more to emit less."

BP launches a carbon footprint calculator in 2004 as a component of its green rebrand. A quarter million people visited BP's calculator in its first year. The term enters the popular lexicon.

Visit BP's calculator today and you're offered an opportunity to purchase carbon offsets. What are offsets? Scant details hide among jargon touting robust industry standards. What matters is, after adding your sins of emission, you may purchase an indulgence. How much would it cost to offset, say, fourteen and a half tonnes of carbon dioxide emissions for the average American? A mere $123! Multiply this by the number of consumers in America, and BP might have us imagine we can fix the crisis with $41 billion per year, an impossible bargain coming in at around one-fifth of 1 percent of the US GDP.

## Carbon-Free Fantasies

The Protestant Reformation famously started when the monk Martin Luther posted ninety-five theses criticizing indulgences on October 31, 1517. The penitent could perform good acts of faith by paying an indulgence and shorten their own or a loved one's time in purgatory. In a phrase attributed to the preacher Johann Tetzel, "As soon as the coin in the coffer rings, the soul from purgatory springs." Supposedly, the funds would build St. Peter's Basilica, but money can be embezzled. Luther attacked Tetzel after the faithful no longer confessed, not to mention the whole ordeal rested on shaky reasoning.

In 1989, an American electric company called Applied Energy Services created the first carbon offset program as part of a plan to build a coal-fired power station. They planted fifty million non-native trees in Guatemala and threatened the ecosystem. Activities upon which locals depended like gathering firewood were criminalized.[1] The company aimed at offsetting the plant's entire lifetime emissions, in the range of sixteen million tonnes of carbon, but an evaluation ten years into the project found less than 2 percent of emissions sequestered.

Carbon offsets are often compared to indulgences, rightly so, especially in voluntary markets. Pay a fee, call it virtue, and enjoy moral absolution. Both the sixteenth-century penitent and the modern beautiful soul likely share in common a desire not to know whether their fantasies match reality. A key difference between indulgences and offsets, though, is that, while people were swindled, they were not otherwise harmed by Tetzel's indulgence campaign. By contrast, the myth of paying a few dollars to plant a tree and erase iniquity very often does mean, in addition to a negligible climate mitigation, the eco-conscious liberal directly expropriates Indigenous peoples of ancestral lands.

In his critique of cultural ideals, Freud described two errors in judgment: illusion and delusion. The former is a belief founded

---

1 Grace Smoot, "The History of Carbon Offsetting," Impactful Ninja, accessed March 25, 2022, impactful.ninja; and Kevin Smith, "Carbon Cop-outs," Transnational Institute, December 1, 2006, tni.org.

upon a wish and the latter a belief that is false. A belief can be one or the other as well as both or neither. If I believe Earth is spherical and average surface temperatures are warming, I experience neither illusion nor delusion. Freud gave a few examples. Columbus's belief he found a route to the East was both illusion and delusion. Aristotle's belief that vermin evolved out of dung was delusion, but it wasn't likely motivated by a wish. If a peasant girl dreams of a prince marrying her, she experiences illusion and likely delusion. The same is true in Freud's estimation for the believer who expects a messiah's arrival.

"In the case of delusions, we emphasize as essential their being in contradiction with reality," Freud wrote. "Thus we call a belief an illusion when a wish-fulfillment is a prominent factor in its motivation, and in doing so we disregard its relations to reality, just as the illusion itself set no store by verification."[2] What made religious beliefs so interesting to Freud was the same thing that interested him in the belief swirling around him and threatening to destroy his Jewish family—namely, the belief that the Aryan race alone is capable of civilization. Racist and religious beliefs are founded on a wish and therefore less susceptible to proofs.

Offsets and self-styled carbon-neutral industries tap into wishes for guiltlessness. In so doing, they offer themselves as illusions. The degree to which they are also delusions will vary.

A mature tree absorbs twenty-two kilograms of carbon dioxide per year. An average American lifestyle annually emits 14.5 $tCO_2$. Were I to scrub those emissions with trees, I could plant about 660 trees today, let them grow to maturity, and ensure they were not cut down anytime in the next century, well after my death, in order to clean up the rest of my life and all emissions so far. This would be a large expense not only in dollars but in time, land use, water, labor, and legal bills.

Nobody seeking indulgences wants to know this. Better to grasp illusions. Reality is a matter of total indifference when the object of purchase is a fantasy of purity. As a result, offsets are often of questionable quality.

Carbon offsets are instruments representing greenhouse gasses in which one offset (or carbon offset credit) equals one tonne of emissions. An instrument of unclear quality is certified by officials

---

2   Sigmund Freud, "The Future of an Illusion," in *The Standard Edition*, vol. XXI, trans. James Strachey (London: Hogarth, 1961), 31.

with varying legitimacy and can be traded for reasons authentic or cynical. If a company emits beyond what regulations permit, it purchases offsets to lower net emissions on paper instead of gross emissions. Activities generating offsets include planting trees, replacement of high-emissions technology with low-emissions technology, or carbon capture and storage. Prices to offset one tonne of greenhouse gasses range from less than two dollars or less per tonne for afforestation, to more than six hundred dollars per tonne for direct air capture of carbon dioxide, so capitalist logic directs money to cheaper projects such as reforestation rather than projects that are most likely to remove emissions from the biosphere.[3]

Recently, a CEO of a lucrative investment firm confessed the $53 million his company earned from offsets had little to no positive effect. In his first foray into offsets markets, the firm sold a restrictive easement to Tennessee barring the company from harvesting an area. Then a carbon project developer suggested they sell credits based on the same land. "We kind of scratched our heads and said, 'Really? You can do this?'" Indeed, they could. Before long, his company pulled money from Chevron to not cut down trees that were already spared. If a company can sell credits based on trees that weren't going to be cut down anyway, say, because they are in a rocky area difficult to harvest, there's money to be made in selling offsets. Offsets purchasers in turn falsely portray themselves as saving trees. This CEO's firm tried to price a real offset (land acquisition plus ongoing surveillance to verify preservation) but found it would cost $60 per tonne. Unfortunately, the average price companies will pay is $3.37.[4] Lower costs mean lower quality. The incentives are all wrong.

A new company called YepYou stretched offsets to the absurd.[5] Billing itself as the "World's First Human Breath Carbon Offset," YepYou offers fees starting at seventeen dollars annually to plant trees to offset exhaled breath. Offsetting your whole lifestyle costs sixty dollars. Customers may even pay for their pet's breath.

---

3 "Carbon Offsetting in North America 101," Ecosystem Marketplace, April 4, 2018, forest-trends.org. See also Pete Smith et al., "Biophysical and Economic Limits to Negative $CO_2$ Emissions," *Nature Climate Change* 6 (2016): 42–50.

4 Ben Elgin, "This Timber Company Sold Millions of Dollars of Useless Carbon Offsets," *Bloomberg*, March 17, 2022, bloomberg.com.

5 "YepYou," accessed March 25, 2022, yepyou.com.

Sustainable or carbon-neutral vacations display this desire to enjoy without knowing too much. Cohica, a "modern travel design agency," sends tourists to tropical paradises with assurances that, as a "TripAdvisor Platinum GreenLeader," travelers' "transportation will be Carbon Offset with the Nature Conservancy."[6] The Nature Conservancy was exposed for counting trees that weren't in danger of being cut. Offsets from this accounting trick go to JPMorgan Chase, BlackRock, and Disney. In 2020, companies bought ninety-three million offsets equal to twenty million cars.[7] So the tourism industry takes advantage of an infrastructure created to help banks and media companies navigate environmental legislation. But all the innocent tourist sees, when they want luxury without guilt, is "Carbon Offset with the Nature Conservancy."

This is the danger of our moment. We're increasingly aware of an ecological crisis, really a polycrisis with too many threats. We are fantasizing subjects, we are subjects who desire, but we are not necessarily subjects who desire to know. The result is what we could call a "purity industrial complex." This complex taps into monetary and libidinal economies of guilt. A purity industrial complex is material (offsets, negative emissions technologies, and green products) and ideological or libidinal (the demand to "educate yourself" on the crisis, sloganeering "science is real," enjoyment of gullibility and self-righteousness covering guilt, and so on). It's a forgiveness machine dispensing absolution from a priestly capitalist climate governance.

Offsets or carbon credits have been played a key role in climate policy since the 1997 Kyoto Protocol and were reaffirmed in Article Six of the Paris Agreement. To understand Article Six, it is helpful to know that Shell's chief climate change advisor David Hone later characterized it as a clear win for the industry: "We have had a process running for four years . . . We can take some credit for the fact that Article 6 is even there at all . . . We put together another straw proposal for the rulebook, and we saw some of that appear in the text."[8]

The governments and organizations certifying offsets are many. The UN Framework Convention on Climate Change tasks host

---

6  "Mexico," Cohica, accessed March 25, 2022, cohicatravel.com.

7  Ben Elgin, "A Top U.S. Seller of Carbon Offsets Starts Investigating Its Own Projects," *Bloomberg*, April 5, 2021, bloomberg.com.

8  Kate Aronoff, "Shell Oil Executive Boasts That His Company Influenced the Paris Agreement," *The Intercept*, December 8, 2018, theintercept .com.

countries with designating authorities who will approve offsets projects, and a third-party auditor should validate. These projects become part of a cap-and-trade system where regulation limits emissions but allows companies to buy and sell credits to meet goals. The EU Emissions Trading System (to be discussed later) is the largest and oldest carbon-credit scheme. Ideally, one credit purchased means one tonne of greenhouse gasses sequestered. In practice, inefficiencies of markets and bureaucracies result in anything from minor, fixable flaws to head-spinning fraud.

A few years ago, one study found 85 percent of carbon emissions reduction programs either do not actually reduce emissions or overstate reduction. Only 2 percent of the offsets programs it examined were highly likely to reduce emissions without overestimation.[9]

Have things changed since then? The existence of sprawling oversight bodies certifying offsets suggests that offsets are here to stay for the time being. Quantifying best practices and gamifying individual guilt hits all the right notes for neoliberal management. The carbon-free fantasy is marketed for individual consumer denial, but as a tool for profit it ingrains denial in business practices. It plays on all-too-gullible or indifferent managers who, when all is said and done, don't particularly care whether what they pay for works. The market finds ways to reap where it does not sow. One airline serves as a case study.

## The World's First Carbon-Neutral Airline

Early in 2020 Delta Air Lines announced it would go carbon neutral by March. The company didn't claim to end emissions but only to "mitigate all emissions," leaving details to customers' imagination: "Because we've committed to carbon neutrality from March 2020 forward, you can feel confident that when you choose to fly Delta, your flight will be carbon neutral."[10] Delta claims its flights are already carbon neutral long before alternatives to jet fuel exist.

---

9  Martin Cames et al., "How Additional Is the Clean Development Mechanism?: Analysis of the Application of Current Tools and Proposed Alternatives," DG CLIMA, March 2016, 11, ec.europa.eu.

10  "Sustainability: Delta Air Lines," accessed March 25, 2022, delta .com.

In one commercial, soft music plays while a woman stands along the shoreline at dusk with her image reflected in the water. Then two young women hike in a green forest, followed by a woman gazing out over a serene lake surrounded by mountains at dawn, but in the water's reflection a wildfire rages. A voiceover asks, "Isn't it a paradox that the love for this world that gets us out in it sometimes leaves behind the things that can harm it?" Another person hops between stones around another lake, while the water below reflects a junk yard. "Flight by flight we broaden our views." A young boy looks at an Earth encased in a toy marble, the atmosphere polluted with yellow and brown smudges as a massive typhoon spins. "And now, flight by flight, we can make a difference." We shouldn't have to choose between seeing the world and saving it, the ad says, and Delta is committed to being "the world's first carbon neutral airline."[11]

Aviation made up 2.4 percent of emissions at the time of the announcement, but this proportion is expected to triple by midcentury.[12] "There is no substitute for the power that travel has to connect people, which our world needs today more than ever before," insisted Delta's CEO. "We know there is no single solution," he claimed, the solution to not burn fuels notwithstanding.[13] Delta's three-point plan leaned on "sustainable aviation fuels," afforestation, and "stakeholder engagement."

Afforestation is a limited tool. Remember that a mature tree absorbs 22 kilograms of carbon dioxide per year. Dividing up a jet's emissions per passenger, each person emits roughly 250 kilograms of carbon dioxide equivalent ($CO_2$-eq) per hour of flight.[14] A round

---

11 "New Campaign Shines Light on Delta's Carbon Neutrality," Delta News Hub, accessed April 29, 2022, news.delta.com.

12 Jeff Overton, "Fact Sheet: The Growth in Greenhouse Gas Emissions from Commercial Aviation," Environmental and Energy Study Institute, October 17, 2019, www.eesi.org.

13 "Delta Commits $1 Billion to Become First Carbon Neutral Airline Globally," Delta News Hub, accessed March 25, 2022, news.delta.com.

14 For clarification, we use two distinct measurements for emissions: carbon dioxide ($CO_2$) and carbon dioxide equivalent ($CO_2$-eq). The latter includes other greenhouse gasses, such as methane, nitrous oxide, and others, some of which produce more warming effect but dissipate quickly. That quick dissipation is why, for instance, you may hear that methane is produces far more short-term warming than carbon dioxide, yet carbon dioxide is the temperature knob for the climate and remains far

trip from New York to London takes nearly fourteen hours, about 3.5 $tCO_2$-eq. That's the annual consumption of about 160 mature trees.

However, Delta didn't plant trees. It bought forests in Indonesia and Cambodia that thrived without Delta's help. The locations meant cheap land. Offsets are a new face of climate colonialism wherein Global North companies buy up property in the Global South and call it good. How many eco-conscious consumers know their green surcharges swindle vulnerable people in the developing world? Racism and ignorance underwrite accounting tricks that swap emissions across ledgers like a game of three-card monte with the biosphere.

At the time of its announcement, Delta was the second-largest emitter among airlines. The industry group International Air Transport Association recently set a goal of 2 percent "sustainable" jet fuel by 2025, up from the 0.002 percent in 2018 that amounted to ten minutes of annual jet fuel use. Sustainable jet fuel means regular jet fuel with some biomass mixed in, not unlike gasoline mixed with corn ethanol. Hydrogen cell-powered planes are in experimental phases, but right now, there's no alternative to jet fuel. If all the space used for jet fuel were converted to state-of-the-art lithium-ion batteries, a 737 could fly perhaps two hundred kilometers.[15]

Delta emitted forty million $tCO_2$-eq in 2018.[16] In 2020, its emissions dropped to nineteen million $tCO_2$-eq due to the pandemic. That year, they'd have hit practically any ambitious reduction target due to fewer flights. In its financial report, Delta claimed to offset only thirteen million tonnes, by purchasing offsets of questionable quality for a couple of dollars each. But in ads, it omitted those

---

more critical. Carbon dioxide equivalent ($CO_2$-eq) is a measure of a gas's potential warming expressed as carbon dioxide. In popular climate change reporting, the difference between these two—$CO_2$ and $CO_2$-eq—is often ignored and left ambiguously as "emissions." I follow the convention of climatologists in using these terms with precision. Annual emissions from all sources are 41.4 $GtCO_2$ or 59 $GtCO_2$-eq. See Intergovernmental Panel on Climate Change, *Sixth Assessment Report*, Working Group III, 2022, 6. Flight emissions from "Aviation," Carbon Independent, February 13, 2022, carbonindependent.org.

15  Jeremy Bogaisky, "The Way Delta Is Going Carbon Neutral Next Month Isn't Good Enough, and CEO Ed Bastian Knows It," *Forbes*, February 15, 2020, forbes.com.

16  "Delta 2020 ESG Report," Delta, 2021, delta.com, 29.

details and said flights were carbon neutral. The millions it spent amounted to barely more than half of 1 percent its typical profits.

Company literature discussed direct air capture of carbon dioxide as part of its approach with a big caveat—namely, it wouldn't actually be using carbon capture. Simply writing about direct air capture's "potential to scale" and "need to progress" allowed Delta to paint itself green. Direct air capture is still in its infancy. The lower end cost by Climeworks, an industry leader, is six hundred dollars per tonne of carbon dioxide. Truly offsetting through direct air capture of Delta's 19.5 million tonnes $CO_2$-eq would cost $12 billion for 2020 (or twice as much in a normal year). For comparison, in the year before the pandemic, Delta's profits were just under $5 billion.

Since actually eliminating its impact would bankrupt the company, you can see why Delta opted for cheap, questionable offsets. But there's apparently nothing too illegal about misleading consumers and investors so long as companies follow best practices expected with capitalist climate governance. In its first sustainability report after its experimental years, Delta claimed: CARBON NEUTRAL.[17]

## Green Petroleum, or Scopes 1, 2, and 3

Nearly two decades after its green rebrand, BP took eleventh place in the Carbon Majors report that found one hundred fossil fuel companies were responsible for 71 percent of annual emissions. Dubbing itself an "Integrated Energy Company," BP pledged to reach net-zero emissions by midcentury. Its sustainability literature says that cities are the real culprit when it comes to emissions.

Per the Carbon Majors report, BP was fifth in historic emissions, behind Saudi Aramco, Chevron, ExxonMobil, and Gazprom. Because BP is a pioneer in capitalist climate governance, it's worth focusing on how it shifted the dodgy discourse on sustainability. How does an oil company claim to go green?

During its early aughts rebrand, the company positioned itself as the conscientious innovator light-years ahead of its oil-consuming customers. I'm indebted to Julie Doyle's research on how BP targeted distinct audiences.[18] In the UK, people were asked whether

---

17    Ibid.
18    Julie Doyle, "Where Has All the Oil Gone? BP Branding and the

they worried about climate change. "I worry because I don't know much about it," replied one, while another said, "It's something that we need to deal with and we need to deal with it today."[19] Blame falls on "I" and "we."

Ads performed authenticity with a man-on-the-street format. Should oil companies do more? "If these guys want to be in business, in fifty years' time," rambled one respondent, "like the responsibility to me and my next generation, and my children and, if they want to be responsible . . . they've got to be investing now in these other alternative, ah, energy sources, I guess."[20] BP is the responsible alternative.

BP boasted it brought solar power to 160 nations at a time when renewables composed 1.6 percent of its investments. In natural gas's heyday as a "transition fuel," BP bragged that four-tenths of its portfolio was gas. Natural gas does emit half the carbon dioxide per unit of electricity compared to coal, but it's still methane turning into carbon dioxide. Ads elided this fact with cheesy jokes: "We've been burning the midnight natural gas," "It's time to think outside the barrel," "Switch on to lower carbon emissions," and "It's time to go on a low-carbon diet."

German ads touted Europe as a renewables leader and praised solar power. "You can't smell it, you can't see it. Is there any better energy than the sun?"[21] UK ads hit the "burning the midnight gas" motif or instructed viewers to "improve your capital."

Playing on American ignorance, ads leaned into basic education with questions like "What is a carbon footprint?" They emphasized BP's activities in the States and promoted stereotypical American values like industriousness or individual moralism. One ad promoting offshore projects in the Gulf of Mexico nostalgically drew on *The Wizard of Oz:* "For finding energy, there's no place like home."

BP's chairman confessed, "Our fundamental objective is to protect and enhance shareholder value."[22] True enough. In many nations, officers of publicly traded corporations have legal obligations to

---

Discursive Elimination of Climate Change," in *Culture, Environment and Ecopolitics*, ed. Nick Heffernan and David Wragg (Newcastle upon Tyne: Cambridge Scholars, 2011), 200–26.

19  Ibid., 227.
20  Ibid., 227–8.
21  Ibid., 234.
22  Ibid., 236.

act on behalf of shareholders' interests, not the biosphere's. Small philanthropic excursions are fine, but changes on the scale necessary would be swiftly punished. That fossil fuel corporations retain large market caps after every new UN climate meeting suggests investors believe net-zero targets are fake. True net zero is as good as zero emissions, but often, in practice, net-zero language massages the urgency and justifies new extraction with gestures at carbon capture later. Perhaps fossil fuel companies aren't deceiving shareholders at all. Everyone sees the intent.

The ads worked. People saw BP as a greener company. The ad firm working the rebrand found BP was viewed as greener (21 percent agreeing) than similar oil companies like Shell (15 percent), Chevron (13 percent), and ExxonMobil (11 percent). Of those familiar with the brands, 49 percent said BP had recently become greener.[23] More than decade later, renewables still barely make up a couple of percentage points of its portfolio.

Yet we know the carbon footprint. We measure our sins by a marketing gimmick. It was based on an earlier, holistic measurement of impacts from an activity or community called an "ecological footprint." It's not right to say BP invented the footprint. Instead, BP took a useful metric, simplified it, and popularized it for a greenwashing campaign. A carbon footprint was simpler. Just tally activities as inputs and see emissions as outputs. A novice consumer turns into an environmentalist by filling out a quick online form.

Travel a bit less. Recycle more. Stop flying. What's the floor? The American carbon footprint of 14.5 $tCO_2$ per capita soars above the global average of 4.8 $tCO_2$.[24] Unsurprisingly, output is higher in oil-producing states. Saudi Arabia averages nineteen tonnes per capita, Kuwait and United Arab Emirates twenty-five, and Qatar a staggering forty-nine. Europe fairs better with states like Portugal, France, and the UK at five or six tonnes per capita.[25] A 2008 MIT study analyzed US lifestyles from children to monks to billionaires. Taking an unhoused person who ate in soup kitchens and slept in shelters as their standard, the study found this individual

23  Gregory Solman, "BP: Coloring Public Opinion?," *Adweek*, January 21, 2008, adweek.com.

24  Ayesha Tandon, "Eradicating 'Extreme Poverty' Would Raise Global Emissions by Less than 1%," Carbon Brief, February 14, 2022, carbonbrief.org.

25  Hannah Ritchie, "Where in the World Do People Emit the Most $CO_2$?," Our World in Data, October 4, 2019, ourworldindata.org.

emits 8.5 $tCO_2$ per year, the equivalent emissions to driving a car fifty thousand kilometers or around twice the global average (for comparison, at the time of the study MIT set the US average at 20 $tCO_2$). The engineer who oversaw the study explained, "Regardless of income, there is a certain floor below which the individual carbon footprint of a person in the U.S. will not drop."[26] Paris Agreement goals require a global average footprint of 1.6 to 2.8 $tCO_2$.[27]

Today, BP underscores its controlling stake in Finite Carbon, a large US developer of forest carbon offsets. From its start in 2013 until BP bought them, Finite Carbon registered seventy million offsets (each offset representing 1 $tCO_2$), or roughly one-sixtieth of BP's total product emissions over that time.[28] But what is BP's actual goal? "Our aim 1 is to be net zero across our entire operations on an absolute basis by 2050 or sooner." Several sleights of hand hide in this target, and BP's deceptive target is, in a way, a mirror image of a flaw in how the public interpreted the Carbon Majors report that simply attributed 71 percent of emissions to a hundred fossil fuel companies. We need to clarify which emissions are counted, and to do that we'll need a set of accounting categories.

Carbon accounting terms Scope 1, Scope 2, and Scope 3 are crucial to understanding what companies and cities actually claim to achieve. Scope 1 emissions are direct, on-site emissions from company vehicles, or any other sources of emissions produced by the business operations of that company. Think of Scope 1 as what you burn through business operations. Scope 2 covers electricity produced offsite at a plant. Think of Scope 2 as what you buy. Scope 3 includes all other emissions from a company's products and supply chain. Think of Scope 3 as all else.

The Carbon Majors report attributed Scope 3 emissions to the oil and gas drillers, a controversial but, in my opinion, fair attribution. BP's goal to drop business operations to net zero is a Scope 1 and 2 goal, which for BP is 55 $MtCO_2$-eq.[29] That's nowhere near the emissions from its oil and gas.

---

26   David Chandler, "Leaving Our Mark," MIT News, April 16, 2008, news.mit.edu.

27   Tandon, "Eradicating 'Extreme Poverty.'"

28   Estimating roughly 4,000 $MtCO_2$-eq from 2013–20, per Richard Heede, "Entity Emissions from Combustion, Venting, Flaring, and Fugitive Methane," Climate Accountability Institute, July 7, 2019, climateaccount ability.org.

29   "BP Sustainability Report 2020," BP, 2021, bp.com, 34.

BP actually does list a goal of becoming completely carbon neutral in Scope 3 by reducing oil and gas production and shifting to blue hydrogen with carbon capture. This doesn't sound realistic to me, and it depends on torturing the definition of Scope 3 to mean upstream production rather than total emissions from their products, but the fine print and the midcentury deadline sure give them time to show off green goals.

Blue hydrogen is a fuel created with natural gas. Fugitive methane leaks and carbon dioxide are inevitable. Critics rightly call it "fossil hydrogen." Industry lobbyists promise they'll capture nine-tenths of emissions, but this hasn't panned out so far. Shell's blue hydrogen plant captures less than half of its emissions and even less if its whole supply chain is factored in.[30] Another option called green hydrogen uses wind and solar power instead of natural gas, so it gives us a true alternative with few to no emissions, but fossil fuel companies would rather push blue hydrogen since it runs on the same methane extracted in their current business model. Blue hydrogen means fossil fuels.

Scopes 1, 2, and 3 give us a language for moral displacement, since one entity's Scope 3 can be another's Scope 1 or 2. To use an analogy from the climatologist Glen Peters, if the crisis were akin to gunning down Mother Earth, who do we blame: the bullet (fossil fuels and companies extracting fossil fuels), the gun (vehicles and power plants), or the person pulling the trigger (you)? I'm not defending fossil fuel companies, but analyses attributing my vehicles emissions to enemies like BP or ExxonMobil's Scope 3 are a choice. To whom do we distribute blame per Freud's two denials? Isn't it interesting that we negate our individual moral culpability and ignore or deny the reality of capitalism's incentives, preferring instead to blame a bad corporation (which, to be clear, deserves blame)?

Total emissions from BP's oil and gas products amount to 550 million tonnes of carbon dioxide equivalent.[31] The 55 $MtCO_2$-eq from operations, the only part BP aims to zero out, are only a tenth of emissions from its oil and gas. No matter! BP can advertise itself as "net zero" (three decades from now) while planning to reduce their footprint by one-tenth of current emissions.

30 "Hydrogen's Hidden Emissions," Global Witness, January 20, 2022, globalwitness.org.

31 Emissions for 2018 were 549 $MtCO_2$-eq per Heede, "Entity Emissions."

When the NewClimate Institute broke down what twenty-five major companies were actually doing with net-zero or carbon-neutral pledges, it found a wide gap between the marketing and reality. These companies were responsible for about 5 percent of global greenhouse gas emissions, and when they advertise net-zero or carbon-neutral goals it looks like they are claiming complete elimination. But the investigation found the pledges would cut emissions by only four-tenths. Companies largely ignored Scope 3 emissions that made up almost nine-tenths of emissions analyzed.[32]

On the one hand, everything I've just described are mitigation tools. On the other, they are weapons in the hands of bad actors to obscure past mistakes and future goals. A fossil fuel company can market itself as "carbon neutral" in Scopes 1 and 2 while still selling petroleum to burn. I suspect we are only a few years away from fossil fuel companies conferring on themselves a net-zero status while, in the fine print, they mean only Scopes 1 and 2, even while emissions from their product persist. Without a broad understanding of Scopes 1, 2, and 3, it will be easy for poor journalism to innocently launder such deception. It's crucial to understand this as more oil companies greenwash themselves with confusing white lies.

## Tesla and the EU Emissions-Trading System

Founded nearly two decades prior, Tesla reported a first full year of profits in 2020. As an electric automaker the company represents a paradox for the climate crisis. Just as the advent of American suburbs constituted the white flight from cities, requiring families to own one or more vehicles, their electric iterations produce new contradictions. Technologies that might well save us down the line present themselves in the guise of absurdly hyped schemes and borderline grifts built by a wealth-hoarding clown with a personality cult.

The moral drive to lower carbon footprints sits right at the surface when people buy electric vehicles. Critics portray EVs as little improvement over combustion since emissions are shifted elsewhere. Cars plug into electricity grids, and the American electricity

---

32 "Corporate Climate Responsibility Monitor 2022," NewClimate Institute, February 7, 2022, newclimate.org.

grids are majority fossil fueled (one-fifth renewables, one-fifth nuclear, one-fifth coal, and two-fifths natural gas). The truth is a mix and changing rapidly. Recent research finds that lifetime emissions of EVs are lower than those of gasoline-powered vehicles, even in regions with dirtier electricity grids. Lifetime emissions of EVs are now lower than comparable gasoline-powered vehicles by half in Europe and the United States and by a third in China and India.[33]

Profitability is a different beast. Tesla put itself into the black by selling carbon credits to polluters in need of offsets. In its first full profitable year, Tesla earned $721 million in profit but took $1.6 billion from credits. Without them Tesla would have posted a loss for the year.

In the United States, fourteen states have regulatory requirements around zero-emissions vehicles (ZEVs). ZEV regulatory schemes incentivize production of EVs by distributing a certain number of credits, which companies producing EVs can sell to other companies that are over budget on emissions. Different countries use different systems. US ZEV credits work differently than Chinese or EU systems. Since Telsa's vehicles are emissions-free at the point of exhaust, it sells credits it receives from various schemes for 100 percent profit. The oldest and largest emissions scheme is the EU Emissions Trading System (ETS). The price fluctuates quite a bit. The price of one tonne carbon dioxide equivalent hovered around twenty-five dollars when Tesla pulled so much revenue from carbon credits.

So for every Telsa purchased, how much is someone else able to emit? Without a more granular analysis of each and every scheme into which Tesla sold credits, the differences between ETS, state schemes incentivizing ZEVs, and other exchanges mean we can only roughly approximate at the risk of gross generalization. If we were using ETS credits when Tesla pulled in $1.6 billion from carbon credit schemes, then Tesla would have theoretically permitted 64 $MtCO_2$. Even before cars plug into coal- and gas-powered grids, selling half a million cars means each Tesla permits someone else to dump a little over 128 $tCO_2$ into the air, equivalent to twenty-eight

---

33  EVs outperform combustion vehicles on emissions by 66 to 69 percent in Europe, 60 to 68 percent in the United States, 37 to 45 percent in China, and 19 to 34 percent in India. See Georg Bieker, "A Global Comparison of the Life-Cycle Greenhouse Gas Emissions of Combustion Engine and Electric Passenger Cars," The International Council on Clean Transportation, July 20, 2021, theicct.org.

years of petroleum-fueled vehicle emissions.[34] You do not purchase a Tesla to lower total emissions so much as to lower *your* emissions.

ETS cuts European emissions by one-twentieth.[35] These numbers change over time but demonstrate the danger that, when you see options to click and pay extra to offset a purchase, you might, instead, do little more than churn out revenue for a clever marketing firm.

## Your Governments Are Lying Through Their Teeth

The Climate Ad Project created a wonderful little cartoon in which a wife walks in on her husband in the middle of an affair. "I would never *cheat* on you," he defends himself. He bought an infidelity offset. "You see, Gus has been cheating on his wife for years," the husband clarifies. "And I paid him not to cheat on his wife for a whole month. So on the whole, the world is a little less 'cheaty.'" The wife stares at the infidelity offset certificate before blurting, "What the fuck?"

As offsets technologies and know-how develop, they might become more effective as both a mitigation tool and a justification for continued fossil fuel extraction. After all that's why Delta and Shell hype carbon capture. Direct air capture is the key example. When a gallon is burned, gasoline results in 8 $kgCO_2$ emissions.[36] At six hundred dollars per tonne, the rate for direct air capture, removing a gallon's worth of gasoline emissions would cost five dollars. I find it difficult, though not impossible, to imagine Americans paying that surcharge without punishing craven policymakers. However, if the direct air capture price came down to one hundred dollars per tonne, bringing the surcharge down to under a dollar per gallon, I can imagine consumers eventually accepting the expense. Klaus Lackner, who heads Arizona State University's Center for Negative Emissions, suggested a similar price range so direct air

---

34  "Greenhouse Gas Emissions from a Typical Passenger Vehicle," EPA, July 21, 2021, epa.gov.

35  Based on cutting 3.8 percent from 2008 to 2016. See Patrick Bayer and Michaël Aklin, "The European Union Emissions Trading System Reduced $CO_2$ Emissions Despite Low Prices," *PNAS* 117, no. 16 (April 6, 2020): 8804–12.

36  "Carbon Dioxide Emissions Coefficients," Environmental Information Administration, November 18, 2021, eia.gov.

capture could allow us to add a surcharge of twenty-two cents per liter of gasoline.[37] Some in the industry are even dreaming of a fifty dollars per tonne future.

The fossil fuel industry would certainly prefer we fixate on this possibility, because it means we needn't strand hundreds of trillions of dollars of fossil fuels. In addition to defending his carbon-heavy lavishness by "buying sustainable jet fuel," Bill Gates, whose carbon footprint is around five hundred times the average American's, boasts of "buying offsets through a company that runs a facility that removes carbon dioxide from the air," saying he is "not aware of anyone who's investing more in direct air capture technologies." This is the paradox of offsets: that which justifies the burning of the planet and delivers wealth to the worst of humanity might well help us later. The left must approach options carefully as direct air capture or other real offsets reach financial viability and scalability.

"A successful new climate policy cannot include any offsets," declared James Hansen, the senior climatologist at the forefront of warning the public for decades. Cap-and-trade sets a floor on emissions; the "perverse effect of the cap-and-trade floor is that altruistic actions become meaningless" when your high-efficiency vehicle lets someone else drive an SUV.[38]

Hansen was deeply critical of the inclusion of offsets in the 1997 Kyoto Protocol. He pointed to an almost comical case of manipulation in which entrepreneurs produced chlorofluorocarbons solely to be destroyed, providing bogus offsets. After Kyoto, emissions continued to rise while offsets proliferated. The same was true after Copenhagen in 2009 and Paris in 2015. Each conference spotlighted targets and nations pledged to do their part, but emissions rose still. "Your governments are lying through their teeth," Hansen concluded. "The truth is that they know their planned approach will not come anywhere near achieving the intended global objectives."[39]

We'd like to rid ourselves of guilt. Barred that option, we buy indulgences. Melancholic guilt comes when an ego eagerly submits to the critical superego—that is, some people want to be told they

---

37    Jeff Tollefson, "Sucking Carbon Dioxide from Air Is Cheaper than Scientists Thought," *Nature* 558, no. 7709 (June 7, 2018): 173.

38    James Hansen, *Storms of My Grandchildren* (New York: Bloomsbury, 2011), 214.

39    Ibid., 183–4.

are wrong and enjoy the security that comes when they no longer have to anxiously wonder. Another type of obsessional neurotic seeks out the analyst's aid in repudiating guilt. Some people know they are not guilty but cannot believe it until an authoritative voice expunges their suspicions. "As far as the patient is concerned this sense of guilt is dumb," Freud said of the guilt converted into physical symptoms. Such guilt "does not tell him he is guilty; he does not feel guilty, he feels ill."[40] We know very well our activity on this planet produces a type of violence that cannot be undone. This knowledge lurks around the edges of consciousness that end up in the most foolish, frantic activity to get clean.

In the end, you purchase a plane ticket, and you check "yes" on the option to donate an extra dollar or so to offset your flight. You know very well it would be better to not fly, but capitalism has not left this option up to you. Carbon dioxide is thus liberated to accomplish what the records of deep time tell us it shall.

---

40  Language on guilt drawn from chapter 5 of Sigmund Freud, "The Ego and the Id," in *The Standard Edition*, vol. XIX, trans. James Strachey (London: Hogarth, 1961).

# PART II

*Sciences*

# 3

# Extinctions

*On some specific day, at some specific hour, the last animal ever will die.*
　　　　　　　　　　　　　　　　　　　　　—Peter Brannen

*In a high meadow, wild bighorn sheep. Their lambs gambol. When you see that gamboling with your own eyes, you'll know something you didn't know before. What will you know? Hard to say, but something like this: whether life means anything or not, joy is real. Life lives, life is living.*
　　　　　　　　　　　　　　　　　　—Kim Stanley Robinson

Far in the future, the sun continues to brighten. Radiation boosts rainfall so that carbon dioxide sinks into the land and washes out to sea. Greenhouse gas levels drop for hundreds of millions of years. Oddly the air is too low on carbon dioxide, yet blisteringly hot from radiation. Eventually, levels drop so low plants cannot photosynthesize.

As food supplies dwindle, the last animals search desperately to feed their malnourished young, but there's nothing to find. Sometime between six and eight hundred million years from now, the last animals starve.

Bacteria inherit Earth for a few hundred million years until they, too, are cooked away. At last the planet is once again barren.

It's unclear how many mass extinctions will happen between today and the last extinction. If history is any guide there will be several, and all will involve fluctuations in carbon dioxide.

## The Precambrian (4.5 Billion Years Ago to 541 Million Years Ago), or the First Seven-Eighths of Earth

Earth is four and half billion years old. For hundreds of millions of years, nothing lived. In deep time, the smallest reactions altered possibilities on the planet. Perhaps near underwater hydrothermal vents, energy expelled from the mantle kickstarted chemosynthetic life. Or perhaps it was lightening in the young atmosphere that arranged chemicals into RNA and DNA proto-life along the path shown by the Miller-Urey experiment. Wherever the origins, microbes appeared more than three and a half billion years ago but nothing more complex. This is the story of subtle changes in Earth's chemistry.

In the next billion years, cyanobacteria evolved the ability to photosynthesize. They sent oxygen into an anoxic biosphere. The Great Oxidation Event began 2.5 billion years ago and injected oxygen into a mostly nitrogen and carbon dioxide atmosphere for five thousand centuries. Before any organism had yet evolved the ability to see or hear, an untold number of forgettable microbes ill-adapted to newly oxygenated conditions likely suffered extinctions. Late in the Proterozoic Eon (2.5 billion to 541 million years ago), early complex life appeared, but Earth was so quiet geologists call a good chunk of this time the "boring billion."

Cellular life specialized as microbes developed organelles and DNA packed into nuclei of new eukaryotic cells. Sexual reproduction was another novelty of this period, though nothing had enough sentience to enjoy it for most of sex's history. After the boring billion concluded eight hundred million years ago (Mya), the first complex organisms arrived. Sponges and other immobile life. They weren't the first to exert changes upon the biosphere, a prize belonging to cyanobacteria, but sponges dutifully busied themselves burying carbon dioxide.

Toward the end of the Proterozoic Eon, during the Cryogenian Period (720–635 Mya), temperatures plummeted and the planet froze into Snowball Earth. Then in the Ediacaran (635–541 Mya) the atmospheric chemistry changed again. Oxygen and carbon dioxide spiked and the warmer climate liquidated life in another mass extinction. With that mix came more twists in deep time.

So concluded the Precambrian, the seven-eighths of Earth's alien prehistory up to half a billion years ago. Then, suddenly, as oxygen and carbon dioxide changed the planet's chemistry, an amazing thing happened.

## The Cambrian (541 to 485 Million Years Ago) and Ordovician (485 to 443 Million Years Ago) Periods: Carbon Dioxide Oscillations

All kinds of organisms multiplied across the oxygenated oceans of the Cambrian Period. A lucky few developed vision, some with five eyes. The earliest nautiloids sniffed and hunted with tentacles. Along the seabed, trilobites scavenged. Trilobites valiantly persisted through the Ordovician and Devonian extinctions before succumbing in the Permian.

The Phanerozoic Eon is our current eon, stretching 541 million years, composed of the old life Paleozoic Era (541–252 Mya), the middle life Mesozoic Era (252–66 Mya), and the new life Cenozoic Era (66 Mya–present). All the "Big Five" extinctions, and many lesser ones, cut down life in the Phanerozoic.

At the dawn of the Ordovician Period, cephalopods patrolled the waters. If their descendants, such as the octopus, squid, and nautilus, are any indication, primal consciousness awakened. The seas of the Ordovician teemed with life but not fish, filled instead with ammonites, sea scorpions, and trilobites. Continents drifted, ripped, and configured anew. Much of what would one day be North America was covered in vast shallows waters, while Australia, India, Arabia, South America, Africa, and Antarctica formed one continent called Gondwana, but even where the continents were dry, life did not yet live on barren land. So much biodiversity thrived in the waters. At the end of the Ordovician, temperatures fell steeply by as much as 10°C. Glaciers formed and oceans retreated.

A critical dynamic to understand: rock weathering changes levels of atmospheric carbon dioxide in Earth's deep time. In the Ordovician, levels of carbon dioxide were very high thanks to rampant volcanic eruptions. High carbon dioxide drives temperatures up, boosting the hydrological cycle. Extra rain grows more acidic due to carbon dioxide, and acidic water, when it falls upon land, chemically reacts with minerals. Sediments flush down mountains and valleys into streams and rivers until they reach the ocean where they fall to the ocean floor as carbonate sediments. Corals absorb and bury some of the matter as calcium carbonate. A smaller scale process happens much the same way with lakes and bogs. Colder waters absorb more carbon dioxide, so changes to either temperature or carbon dioxide amplify the other's effect. Rock weathering is Earth's thermostat if given enough time. Today, weathering draws carbon dioxide out of the atmosphere at a little over a tenth of one part per million per year. When volcanic carbon dioxide injection outpaces weathering, the result is a hothouse atmosphere and acidic oceans. When weathering outpaces volcanic carbon dioxide injection, the glaciers advance.

At the end of the Ordovician, young life-forms unlike anything to ever exist faced both catastrophes. First the heat, then, when the volcanism calmed, a deep freeze. Over a couple million years as greenhouse gases levels oscillated, as much as 85 percent of life perished.

## Devonian Period (419 to 359 Million Years Ago): The First Trees

Life returned. After the advance and retreat of life in the Silurian Period, the world of the Devonian exploded with fish and insects. Massive armored placoderms ruled the seas. A few of the first tetrapods, the ancestors of four-limbed land animals, pulled themselves onto the shore to catch insects. Some had six or eight finger-like digits, while the ones who survived had five.

The star of the Devonian, though, wasn't an animal. Unlike the barren continents of the Ordovician, small flora proliferated in the Devonian. One plant, *Archaeopteris*, developed a deep-root system that allowed it to rise thirty meters into the sky. The first tree.

With their roots tearing though the land, trees created soil. Liberated nitrogen washed into streams and out to sea, feeding the algae spreading at the surface and clearing out oxygen from the water below. Or perhaps volcanic activity injecting carbon dioxide raised sea level and overwhelmed life struggling to survive along the continental coasts. Whatever the cause of the deadly Kellwasser Event (374 Mya), the climate toggled to extinction conditions again over millions of years. By the conclusion of the Devonian, the Hangenberg Event (359 Mya) wiped out stragglers. The trees advanced inland and breathed carbon dioxide until temperatures dropped 20°C. The remains turned into shale and natural gas.

In the Carboniferous Period (359–299 Mya), flora spread over the continents. When the plants died, starting in the Carboniferous but continuing forever, Earth locked up the carbon safely as coal.

## Permian Period (299 to 252 Million Years Ago): The Great Dying

The Paleozoic world was brought to an end in the greatest loss of life before or after: the Great Dying. At the end of the Permian Period, carbon dioxide rose so steeply and suddenly that even plant species craving carbon dioxide couldn't adapt, leaving a ten-million-year gap in coal deposits.

This was the time of Pangea, the supercontinent surrounded by the super-ocean Panthalassa. As with all mass extinctions, the details remain somewhat hazy. Siberia erupted with flood basalts—not normal, singular volcanic eruptions but continuous flows of lava spilling from Earth's mantle—and, with them, sulfur dioxide and carbon dioxide. Lava lit up coal deposits and assaulted the atmosphere with volleys of greenhouse gas. Pangea's arrangement, a supercontinent with less coastal space, meant less shoreline for rock weathering and more carbon dioxide concentrating unabated. Heat disrupted the ocean circulation current, so fewer storms over the supercontinent meant less rainfall and ever less weathering. Changes to ocean circulation withheld oxygen from the deep ocean and killed large life while bacteria thrived and sent up sulfur dioxide. Perhaps the gas killed land animals on its way up to shred the ozone. An overload of radiation passed through sulfur dioxide skies, which, due to the change in chemistry, might have been tinted visibly green.

It isn't just temperature change that matters. The rate of change is critical. Plants and animals need time to adapt. The minimum threshold for mass extinctions is greater than 5.2°C temperature change (warming or cooling) at a rate of greater than 10°C per million years. A 10°C average temperature at the beginning of the Permian turned to 32°C by the end. Too much, too quickly.

As much as 95 percent of life ended in the Great Dying. Casualties turned into coal and oil. It took only seventy-five thousand years because carbon dioxide rose faster than at any point in Earth's history until the twenty-first century, when carbon dioxide concentration rose about a hundred times faster than in the Permian.

## Triassic Period (252 to 201 Million Years Ago), or the End of Pangea

Life returned again in the Triassic Period and thrived until the boundary with the Jurassic. Among the Triassic arrivals are dinosaurs and the first mammals. They thrived in the newly wet conditions brought by the Carnian Pluvial Event, the humid greening of the land.

Pangea pulled apart and created the Atlantic Ocean. The rift turned into eruptions. Flood basalts threatened the young dinosaurs and mammals with lava. Lava underwater and on land dumped greenhouse gasses into the atmosphere. Pores on plants grew smaller in the Triassic, since carbon dioxide was easy to come by, but the same gasses turned oceans acidic. The modern coral reef started here, but reefs cannot survive when carbon dioxide levels rise much above 450 parts per million (ppm). Reefs are such crucial hosts to ocean life, and a system fine-tuned to certain levels of oxygen, acidity, and sunlight cannot relocate when rising carbon dioxide acidifies the water or raises sea level, dimming the sunlight for the reef below. Warming worked slowly over a million years at the end of the Triassic, but temperatures ticked up again. Three-quarters of species died out at the Triassic-Jurassic boundary.

## Cretaceous Period (145 and 66 Million Years Ago) and Only Exception to the Rule

The age of the Tyrannosaurus ended with a famous last scene we all know. At some specific moment on some early summer day, while water lilies were blooming, a ten-kilometer-wide asteroid pushed through the atmosphere at more than twenty kilometers per second and touched down upon the Yucatán Peninsula. Animals close enough to see—that is, if nervous systems transmitted the blinding light to their brains before roasting—were destroyed by a hundred-million-megaton blast. A few hours of hell followed by a long impact winter. The blast from the bomb criminally dropped on Hiroshima was seven million times smaller than the end-Cretaceous impactor.

All other big extinctions correlated so closely with carbon dioxide rises or drops that it makes sense to think of them as caused by carbon dioxide fluctuations. The end-Cretaceous extinction is the sole exception to this rule, but it is not entirely free of the green-house effect.

Since the 1980s, the conventional theory for the end of the dinosaurs has been an impactor, but another theory posits carbon dioxide injection from flood basalts covering parts of the Indian subcontinent in several kilometers of lava as it crashed into Asia. Still another theory supports some combination wherein global earthquakes from impact triggered volcanism elsewhere. Whatever the relation in this most atypical extinction there was still a green-house effect, and, whatever the sequence at the end-Cretaceous, temperatures rose 6°C.

The land around the impactor filled in and left little trace of a crater beyond an arc of terrain buckled from impact, much like ripples when a stone lands in a pond. Freshwater seeped into these limestone ripples. Some of the limestone collapsed into the empty caverns below, creating sinkholes to access freshwater that future inhabitants will call cenotes. The Mayan civilization built itself in a perfect half-circle at the edge of the Yucatán Peninsula by settling around these vital cenotes, the aftermath of the impactor. After their civilization collapsed, surviving descendants were murdered by the European forefathers of fossil capitalism. Some sixty-six million years after impact in Chicxulub Puerto, the very center of the catastrophe was memorialized with a tacky slab of concrete. On the monument, two dinosaur skeletons are painted, and below them a few lines are scribbled: "AN ENORMOUS ASTEROID 10 KM IN DIAMETER HIT IN THIS LOCATION."

Sea level was much higher throughout the Cretaceous and covered much of North America in shallow water. Creatures died and their calcium carbonate exoskeletons accumulated on the continental floor, which would become the American South. When the seas eventually retreated and exposed fertile soil, this region drew slaveholders to reap profits from cash crops, brutalizing humans as a novel mode of capitalist production set off a dangerous chain reaction.

## Paleocene-Eocene Thermal Maximum, or the Last Time There Were No Ice Caps

Thus began the Cenozoic Era (66 Mya–present). Temperatures declined a few degrees over a few million years of the Paleocene

Epoch (66–56 Mya) then suddenly halted. Volcanic activity started again and injected so much carbon dioxide that the juncture between the Paleocene Epoch and the Eocene Epoch (56–34 Mya) takes a special name: the Paleocene-Eocene Thermal Maximum. At this moment, some fifty-six million years ago, carbon dioxide soared to 1,400 ppm and temperatures rose 5°C. Outside the Big Five extinctions, this minor extinction was the only time temperatures changed faster than a rate of 10°C per million years. Today, our climate's pace of change is orders of magnitude faster.

The hot Eocene Epoch was the last time there were no ice caps at the poles. Carbon dioxide hit 800 ppm. Oceans were seventy-five meters higher. Crocodiles sunned themselves near Arctic palm trees.

In the Oligocene Epoch (34–23 Mya) and Miocene Epoch (23–5 Mya), rock weathering reworked Earth's chemistry while volcanoes mercifully quieted, and carbon dioxide dropped to near 450 ppm. At that point, starting thirty-four million years ago the oceans retreated as a new semipermanent feature emerged on Earth's surface: the Antarctic ice cap. First, only the high mountains were coated, but over tens of millions of years the temperature dropped and permanent glaciers filled the continent.

Carbon dioxide eventually dropped to what we now call pre-industrial levels, but first, there was one final spike when concentration once more sailed past 400 ppm, in the Pliocene Epoch (5.3–2.6 Mya). Temperatures hovered as high as 3°C above pre-industrial levels. That was the last time there was anything comparable to the amount of carbon dioxide in the air in the twenty-first century. Oceans jumped fifteen to twenty-five meters.

## The Quaternary Period (2.6 Million Years Ago to Present): The Ice Age

Over the rest of Earth's history to date, the Quaternary, ice advanced and retreated more than fifty times. But carbon dioxide injection stabilized, so three other dynamics called Milankovitch cycles, usually too small to overcome the greenhouse effect, forced Earth into warmer and cooler periods: the 41,000-year cycle of obliquity or tilting of the axis, the 26,000-year precessional cycles in which Earth's axis wobbles in a circle and the 100,000-year cycle of oscillation between more elliptical and more circular orbits. In common usage, people speak of the cool periods as ice ages, but paleontologists call the entire Quaternary an ice age, right up to the twenty-first century. Earth's cycles have pushed the surface into cool glacials and warmer interglacial periods such as now.

The Quaternary is composed of two epochs: the Pleistocene Epoch (2.6 mya–11,700 years ago) and the Holocene Epoch (11,700 years–present). Some argue we'll need a third epoch called Anthropocene.

All members of the *Homo* genus descend from Australopithecus, which first appeared four million years ago. A new hominid lineage branched off some two million years ago in the Pleistocene. In Africa *Homo erectus* walked upright, and different subspecies split off and made their way to Asia and Europe for well over a million years. *Homo heidelbergensis* was our ancestor. They evolved in southern Africa, but when the orbital cycle altered climate suitability between 430,000 to 400,000 years ago, *H. heidelbergensis* advanced north to Europe. They built fires to keep warm and eventually became Neanderthal. Another branch of *H. heidelbergensis* remained isolated in southern Africa until the orbital cycle changed habitat suitability again between 300,000 to 200,000 years ago. The group genetically isolated by climate conditions could now disperse into the world as it speciated. *Homo sapiens.*

The species emerged out of Africa in several waves. They evolved, mated with archaic humans, and split off again, all while adapting strategies to survive climate changes. In an unusually hot time 125,000 years ago, the Eemian interglacial period sent temperatures rising for the last time before the twentieth century. The climate briefly warmed 2 to 3°C above pre-industrial levels, or in the range of what we may see later this century. Oceans pulsed five to ten meters higher. Aside from two smaller temperature spikes 240,000 and 216,000 years ago, until recent years the Eemian was the only time humans ever experienced temperatures above the pre-industrial norm.

The exact age of *H. sapiens* is unclear as new evidence emerges. *H. sapiens* bone fragments dating to 300,000 years ago have been found in northern Africa. Between 100,000 and 40,000 years, fossils show a mix of modern and archaic features. Earliest *H. sapiens* fossils outside Africa appear in the Middle East, dating to 194,000 to 177,000 years ago. *H. sapiens* penetrated Asia and reached China 80,000 years ago at the latest. One massive migration out of Africa 60,000 to 50,000 years ago was made possible by the lower sea levels of the post-Eemian glacial period. At Qafzeh and Skhul, near Haifa, Israel, *H. sapiens* and possibly Neanderthals lived in close proximity in caves dotting the mountains. Skeletons were painted with red ochre pigments indicating burial rituals. They were mourning their dead.

On an archeological dig at Tel Akko, I took a day trip to visit the excavated site at Skhul. I won't ever forget a hike up the hills to small caves in which our ancestors sought shelter, congregating with other more or less modern humans. A sacred space for my disenchanted soul.

Disaster struck seven hundred centuries ago. Fewer than ten thousand individual humans were left alive. Some estimates pin the bottleneck at one thousand reproducing pairs. For a while the catastrophe was attributed to the eruption of Sumatra's Mount Toba seventy-four thousand years ago. Aerosols rose high in the atmosphere and blocked sunlight. Temperatures dropped and humans almost died out. However, this theory has recently come under scrutiny, in part due to the possibility of several population bottlenecks in this period that do not neatly line up with the Mount Toba eruption. Populations declined, the temperature dropped, and the volcano had disastrous regional impacts in Southeast Asia, but the causes for bottlenecks remain subject to investigation.

Populations recovered over many years. Estimates range from twenty to five hundred generations. Humans developed spoken language 50,000 to 100,000 years ago. They reached Australia 65,000 years ago and North America 24,000 years ago, maybe more. Wherever they arrived, hunters with large brains extinguished prey. Gone were famous beasts like the mammoth and forgotten were sights like four-meter giant sloths. Neanderthals couldn't keep up and died out 30,000 years ago.

Right around that time, three hundred centuries ago, there was an explosion of symbolic expression. Our ancestors painted the hunt on the walls of Chauvet-Pont d'Arc Cave. Images depicted motion, a sophisticated skill. Elsewhere, a genetically isolated group of Cro-Magnon built mass graves, as did other *H. sapiens*. Symbolic expression spread far and wide, even if archaeological discoveries are prejudiced by where European university departments search. The emergence of artistic expression remains an enigma, as do social implications of burial rituals. In the Russian Sunghir, at a site dated twenty-eight thousand years ago, two children were buried head-to-head, their skeletons adorned with red ochre pigments and covered in ivory beads, mammoth tusks, daggers, fox teeth, bracelets, and pendants. Depending on interpretations, these were either royal burials, early victims of human child sacrifice, or simply a demonstration of increasingly elaborate decoration of the deceased. Humans made Venus figurines, small statues with little in the way of heads or appendages but large breasts and carefully sculpted vulvas, often with swollen abdomens possibly indicating pregnancy. Depending on interpretations, these were either fertility goddesses, pornography, dolls, ancestors, or simply art. In the Czech Republic at a site dated twenty-six thousand years ago, figurines were made from fired clay. At the Russian site Yudinovo, "mammoth houses" constructed from the remains of the great beasts seem to be monuments to the hunt. At Göbekli Tepe, humans

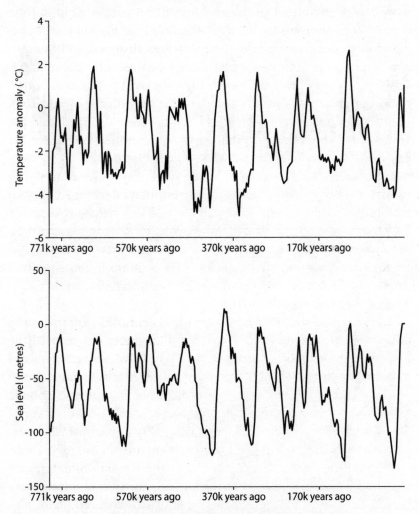

Figure 3.1. Milankovitch Cycles Affect Temperature and Sea Level

erected stones to create a circular monument nearly a hundred centuries ago, perhaps the first temple.

While humans survived and managed to make meaning, the cyclical wobble of the planet pulled our ancestors out of the last "ice age," or glacial period, and into the current interglacial period. As we saw above, Earth's Milankovitch cycles alter eccentricity, obliquity, and precession over tens of thousands of years. Absent so much volcanic activity injecting carbon dioxide, these subtle changes are more than enough to drive the climate between glacial and interglacial periods. Colder climates mean colder waters absorb extra carbon dioxide and draw temperatures even lower.

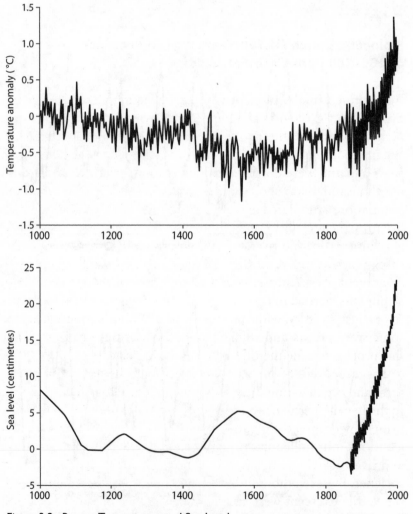

Figure 3.2. Recent Temperature and Sea Level

Starting around fourteen thousand years ago, sea level increased four or five meters per century for several centuries and temperatures rose 5°C. For the last 800,000 years carbon dioxide hovered between 180 and 280 ppm and never deviated outside that range until the twentieth century. The new global average temperature at the beginning of the Holocene is what we refer to as the pre-industrial average, and all of civilization has taken place within a degree of this interglacial temperature range. All contemporary religions, philosophies, and values were born after—and thus can little conceive a world without—the class hierarchies enabled by agricultural surplus made possible by one configuration in Milankovitch cycles.

## Holocene Epoch (11,700 Years Ago to Present), or 180–280 ppm Carbon Dioxide

Agriculture was absent in the Pleistocene but a necessity in the Holocene. What geologists call the beginning of the Holocene Epoch, archaeologists call the beginning of the Neolithic Period. Humans began modifying wild plants as early as eleven or twelve thousand years ago and were domesticating the "founder crops" by eight to ten thousand years ago. Starting in the Fertile Crescent, growing populations stimulated an agricultural revolution in grains, which led to higher fertility. Communities in Southeast Asia developed agriculture too, and over the next few thousand years agriculture spread or was independently developed. Five thousand years ago, an anomalous rise in methane corresponded to changes in land use as humans learned to flood lowlands and irrigate crops. According to one just-so story, with surpluses of food, labor divided such that some were priests and soldiers among majorities of farmers. But this process was neither linear nor immediate.

The record tells us our ancestors experimented with many agricultural and organizational styles, some more egalitarian than others. In the Levant, permanent settlements in the Natufian period existed prior to widespread crop and animal domestication, and there was a curious four millennia gap between the earliest grain cultivation and livestock domestication and the first truly agrarian-pastoral societies. By seven thousand years ago, hundreds of communities in the Fertile Crescent were cultivating crops. The nature of the relationship between crops and hierarchical societies is contested, but over the first few thousand years of agricultural development, our ancestors built the earliest towns, like the settlement at Çatalhöyük some nine millennia ago with a population eventually as high as ten thousand. The early agrarian states along the Tigris and Euphrates rivers, such as Uruk, built on the backs of slave labor we typically think of as the effect of crop domestication, did not dominate southern Mesopotamia until five to six thousand years ago.

In the unusually stable interglacial period, carbon dioxide never budged above 280 ppm. Unlike the end of the Pleistocene, when the seas rose a meter every couple of decades for several centuries, Holocene shorelines eventually stabilized. Trade required a memory technology. Fifty-plus centuries ago, the Sumerians developed cuneiform. The cumbersome written language required scribes to commit hundreds of signs to memory. They recorded their trades conducted along the Euphrates and over land. One of the earliest common symbols means "female slave of foreign origin," a testament to Uruk's class society emerging alongside writing and surplus labor. Thirty centuries ago, seafaring Phoenicians used their alphabet to record trades throughout the Mediterranean settlements made possible by stable carbon dioxide that had fixed the sea level a millennium prior. As a side effect of this technology humans eventually recorded oral traditions as written texts: the Vedas, the Enuma Elish, and the Epic of Gilgamesh.

We traded and consumed energy from plants and animals, exposed to small fluctuations in the climate we couldn't possibly understand. Once Europe began its exploration and genocides in the New World while dealing with a reformation and wars at home, the Little Ice Age dropped global temperatures a fraction of a degree for several centuries. The dates of the Little Ice Age are contested, but what seems clear is a quarter-degree temperature drop in the thirteenth century (likely from reduced solar activity and perhaps increased volcanic activity) was followed by a plummet of an additional half degree when Europeans arrived in the Americas (a popular theory being that the Indigenous genocide resulted in forest growth sinking carbon dioxide), a drop from which the world did not recover until the nineteenth century. During the cold, famines and bread riots and angry peasants meant political instability and scapegoating. Europeans shrank from malnutrition and blamed Jews for plague and accused women of witchcraft. Hostilities boiled into the Thirty Years' War, a dynastic struggle in the guise of religious war crossed with emerging economic competition, all of which, along with hunger, battle, and social unrest, contributed to the general crisis of the seventeenth century. The war concluded with the Peace of Westphalia weakening the Holy Roman Empire as state sovereignty took shape, though not before Puritans and other sects sought refuge by joining in the colonization of the Americas. Protestantism turned faith into a matter of individual conviction, starting a countdown to the death of God at the precise moment a new, post-feudal mode of production distributed coats and food produced by wage labor with varying success on a market. Enslavement, land enclosure, and novel financial techniques and entities like the joint-stock company supercharged a heretofore unimaginable level of surplus exploitation as the emerging capitalist mode of production created new social architectures practically begging for an explosion of cheap energy. Colonizers in the Americas faced harsh winters, but their markets and beliefs consumed every square meter of the world.[1]

---

1 I strove to limit and condense footnotes for the sake of narrative and aesthetics in this chapter, but many sources contributed to my writing. For the relationship between carbon dioxide and extinctions I'm indebted to Peter Brannen, *The Ends of the World* (New York: Ecco, 2017) and Peter Ward, *Under a Green Sky* (New York: Smithsonian Books, 2007). Data on extinction temperatures and rates of change from Song Haijun, "Thresholds of Temperature Change for Mass Extinctions," *Nature Communications* 12, no. 4694 (August 4, 2021) and Paul Voosen, "A 500-million-year Survey of Earth's Climate Reveals Dire Warning for Humanity," *Science*, May 22, 2019, science.org. Figures 3.1 and 3.2 of temperatures experienced by humans from "Global Temperature Record," 2° Institute, accessed February 2, 2021, sealevels.org. Modeling on Milankovitch cycles driving hominin speciation and geographical spread from Axel Timmermann et al., "Climate Effects on Archaic Human Habitats and Species Successions," *Nature*, April 13, 2022, 1–7. Some data on human origins from Brian Handwerk, "An Evolutionary Timeline of Homo Sapiens," *Smithsonian Magazine*, February 2, 2021, www.smithsonianmag.com. Little Ice Age temperature from Intergovernmental Panel on Climate Change, *Sixth Assessment Report*, Working Group I, 2021, 317. Timescales per the Geological Society of America. Miscellaneous climate, agriculture, and human population information from Patrick Wyman, "Uruk and the Emergence of Civilization, *Perspectives*, Feburary 18, 2021, patrickwyman.substack.com, Hansen, *Storms of My Grandchildren*; Vaclav Smil, *Growth* (Cambridge: MIT Press, 2020); David Graeber and David Wengrow, *The Dawn of Everything* (New York: Farrar, Straus and Giroux, 2021); James C. Scott, *Against the Grain* (New Haven: Yale University Press, 2017); and Matt Christman, various podcasts.

Throughout the entire existence of *H. sapiens*, the climate was only ever warmer in the Eemian interglacial period, and then only barely. But in the late eighteenth century, James Watt invented a steam engine powered by coal. All heretofore existing civilization had drawn power from animals or biomass such as wood. Not so anymore. Before long, we adapted to the internal combustion engine.

Capitalist expansion, settler-colonial racism, technological development, and electrification drew on fossil fuels as a one-time injection of cheap energy. In early days of the population boom, prophets of doom like Thomas Malthus predicted mass starvation. But at the start of the twentieth century, the chemist Fritz Haber discovered how to fix nitrogen to fertilize land, which in conjunction with the agricultural Green Revolution later in the century, made it possible to feed billions extra and, in the process, change Earth chemistry in yet stranger ways. Sometimes, we engineer our way out of a problem. Sometimes we don't. Around the same time the Haber process came online, another chemist, Svante Arrhenius, figured out rising carbon dioxide injected by coal would raise temperatures. It was nearly a century before the public learned there was a problem.

# 4

# Carbon Dioxide

*What the climate needs to avoid collapse is a contraction in human-
ity's use of resources; what our economic model demands to avoid
collapse is unfettered expansion. Only one of these sets of rules can
be changed, and it's not the laws of nature.*

—Naomi Klein

*In its relation to desire, reality appears only as marginal.*

—Jaques Lacan

On an otherwise unremarkable Wednesday in November 2006,
the US Supreme Court hears arguments in *Massachusetts v. the
Environmental Protection Agency*. Plaintiffs want the EPA to regulate greenhouse gasses. The Clean Air Act requires it, they argue.

The lawyer for the plaintiffs, James Milkey, makes his case for
the EPA's role in regulation. Justice Antonin Scalia, a self-confident
blowhard, interjects. "If we fill this room with carbon dioxide,
it could be an air pollutant that endangers health. But I always
thought an air pollutant was something different from a stratospheric pollutant," Scalia rambles in confusion and immediately
repeats his mistake. "Your assertion is that after the pollutant leaves
the air and goes up into the stratosphere it is contributing to global
warming."

After the second time, Milkey feels a need to clarify. "Respect-
fully, Your Honor, it is not the stratosphere. It's the troposphere."

"Troposphere, whatever. I told you before I'm not a scientist,"
Scalia confesses as the room erupts in laughter. It's funny to know
nothing. "That's why I don't want to have to deal with global
warming, to tell you the truth."[1]

---

1 *Massachusetts et al. v. Environmental Protection Agency*, 549 US
497 (2007).

## On the Need to Address Knowledge Gaps

Starting in the late eighteenth century, the capitalist mode of production incentivized rapid extraction and combustion of coal that had safely sequestered carbon as far back as the Carboniferous and Devonian Periods. Between industrialization and when I write this sentence in 2022, global average surface temperature rose 1.3°C. Land temperature rises 50 percent faster than the average, so land temperature rose about 1.9°C while oceans rose 0.9°C.[2] The rate of warming is now at least 0.18°C per decade. Heat is held in the lowest level of the atmosphere, called the troposphere. A jet at cruising altitude in the lower stratosphere flies above the whole problem. Heating will continue until emissions reach net zero.

Journalists take one of two broad paths in reporting climate change. One conveys the basics in layman's terms, generally making an effort to describe a fair variety of scenarios. This option usually sprinkles in one to three sciency-sounding fun facts and avoids technicalities. The other option, which unfortunately draws more clicks, is to paint vivid horrors and underscore worst-case scenarios. Both options leave gaps anyone is perfectly capable of understanding if given a shot. This chapter closes those gaps.[3]

How will effects of 2°C warming differ from the more drastic consequences of 4°C? Climatologists caution against viewing a temperature, point on a carbon budget, or deadline year as a hard threshold. It's like asking whether cancer sets in after 100,000 cigarettes or 150,000. The IPCC acknowledges massive uncertainties around tipping points (see footnote for a list of tipping points beyond the big cataclysms detailed at the end of this chapter).[4] We

---

2  Zeke Hausfather, "State of the Climate: How the World Warmed in 2019," Carbon Brief, January 20, 2020, carbonbrief.org; and Zeke Hausfather, "Zeke Hausfather: State of Climate Science, Energy Systems, Post COP26, Tipping Points, Tail Risks," interview by Benjamin Yeoh, *Ben Yeoh Chats*, November 22, 2021, podcasts.apple.com.

3  I wish to acknowledge this chapter is loosely modeled on James Hansen's excellent introduction to basic climate science. See James Hansen, "Climate Change in a Nutshell," Columbia University, December 18, 2018, columbia.edu.

4  For some reason, it's popular to say the IPCC doesn't acknowledge tipping points or doesn't acknowledge them sufficiently, but, in fact, Working Group I strongly warns of nonlinear tipping points in the paleoclimate record, such as when a change of merely one degree melted the

can only model so much. We expect 15 percent crop loss at 2°C, but, hypothetically, say we discover crucial flora on which bees depend dry up a tenth of a degree earlier: governments are unable to adapt to famines; mass insurrections destabilize supply chains; depression, fascist reactions, and war follow; what else? Or in a more quotidian malaise, imagine social tipping points when insurance companies go under and people can no longer protect houses that might not survive a thirty-year mortgage. When people's homes (primary investment instrument) decline in value, their property taxes drop, school funding and infrastructure suffer while police budgets balloon, property is devalued further, residents leave, and the community suffers.[5]

Divining the future with exactitude is a fool's errand. Our foreknowledge tells us consequences will be dire and likely ignored until it's too late. However, the basic physics are less mysterious and, to say it all, almost tragically straightforward.

## Emissions at a Glance

Emissions are a sin of wealth. The top tenth of households contributes nearly half of emissions; the top 1 percent alone are responsible for 16.9 percent of emissions, while the poorest half of humanity contributes only 11.5 percent.[6] Fault and suffering are mismatched.

---

glaciers or slowed the Atlantic Meridional, overturning circulation. Similarly slight changes might do far worse with extreme weather, droughts, and fires. See Intergovernmental Panel on Climate Change, *Sixth Assessment Report*, Working Group I, 2021, 106, 1148. On more than a hundred pages, Working Group II warns of tipping points including sudden Arctic permafrost or tropical forest loss, fisheries collapse, extinction, loss of livelihoods particularly for Indigenous communities, and need for communities to relocate. Intergovernmental Panel on Climate Change, *Sixth Assessment Report*, Working Group II, 14–20, 14–69, 16–118. Working Group III warns of natural carbon sinks switching to emitters and castigates banks for failing to disclose how their portfolios are exposed to unexpected tipping points. See Intergovernmental Panel on Climate Change, *Sixth Assessment Report*, Working Group III, 2022, 4–104, 15–56.

5   This example of a social tipping point is inspired by the climatologist Andrew Dessler.

6   Lucas Chancel, "Global Carbon Inequality over 1990–2019," *Nature Sustainability* 5 (September 29, 2022): 931–38.

Globally, all human activity emits approximately 41.4 gigatonnes of carbon dioxide ($GtCO_2$) per year. Historically, the United States is the largest carbon dioxide contributor and responsible for fully one-fifth of emissions at 509 $GtCO_2$, followed by China at 11 percent, Russia at 5 percent, and Indonesia at 4 percent. Germany and the UK contribute 4 and 3 percent, respectively.[7]

Since carbon dioxide corresponds linearly to temperature, the historical percentage indicates the warming each country contributed. China emits a quarter of all carbon dioxide today, more than any other. India is on the rise at one-fourteenth of today's emissions. The United States emits one-seventh and the EU one-tenth. But, historically, the United States alone is responsible for over 0.2°C of global warming, whereas China has only contributed half that. A US resident is second only to a Canadian in per capita cumulative emissions.[8]

Of all greenhouse gas emissions worldwide (2022, latest data available), 17.5 percent comes from energy use in buildings, 24.2 percent from energy use in industry, 16.2 percent from transportation, 18.4 percent from agriculture, forestry, and other land use, 3.2 from waste, 5.2 from direct industrial processes, and another 15.3 percent from unallocated fuel, combustion, fugitive emissions, and energy in agriculture. Focusing only on 41.4 $GtCO_2$, 35.1 $GtCO_2$ come from fossil fuels (15.1 coal, 12.1 oil, and 7.9 gas), another 3.9 come from agriculture and land use, and the last 2.4 come from industry and waste.[9]

You probably have your head around what transportation and electricity entail, but "industry" is a bit of a mystery box to most people. Industry means most activities aside from energy or agriculture. It includes manufacturing, mining, construction, and waste processing. It's the physical economy. If we clean up our energy grid and electrify transportation, it will not resolve the different problems of industry emissions. Of the millions of different activities encompassed by industry, the vast majority of industrial emissions (two-thirds) are due to steel, cement, and commodity chemicals

7   Simon Evans, "Analysis," Carbon Brief, May 10, 2021, carbonbrief .org.

8   Ritchie and Roser, "$CO_2$ and Greenhouse Gas Emissions"; and Evans, "Analysis."

9   Data for Figure 4.1 from Friedlingstein et al., "Global Carbon Budget 2022"; and Hannah Ritchie and Max Roser, "Emissions by Sector," Our World in Data, 2022, ourworldindata.org.

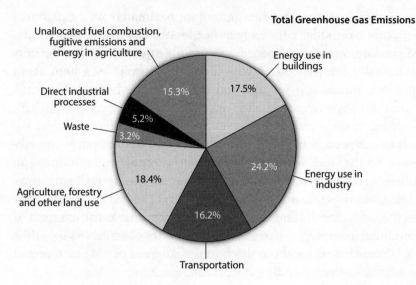

**Total Greenhouse Gas Emissions**

Unallocated fuel combustion, fugitive emissions and energy in agriculture — 15.3%

Direct industrial processes — 5.2%

Waste — 3.2%

Agriculture, forestry and other land use — 18.4%

Transportation — 16.2%

Energy use in buildings — 17.5%

Energy use in industry — 24.2%

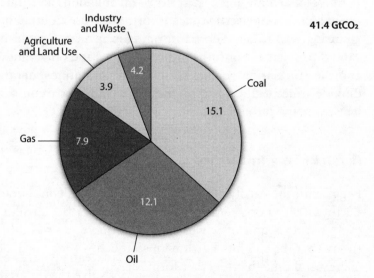

41.4 GtCO₂

Industry and Waste — 4.2

Agriculture and Land Use — 3.9

Gas — 7.9

Oil — 12.1

Coal — 15.1

Figure 4.1. Emissions

(for plastics and fertilizers). Waste processing is an example of the remaining third, and we mostly know how to reduce methane emissions, but creating cement and steel without carbon dioxide emissions is much harder. For example, steel is made from iron ore (iron oxide, or rust) heated up to tremendous temperatures. The heat is often generated from coal, but coking coal is also used to strip oxygen away from the iron oxide, which becomes carbon dioxide. In other words, steel production produces carbon dioxide in two ways, both from burning coal for heat and from the chemical reaction of carbon with oxygen in the iron oxide (the latter emissions from chemical reaction, not combustion, are called process emissions). We use a lot of steel too, about a half tonne per person per year globally. If we can transition to using hydrogen to strip away the oxygen instead, then in conjunction with clean electricity we can create green steel.[10]

Decarbonizing industry is difficult and requires public investment. In the week I write this, the Biden administration is directing the federal government to use building materials derived from low-carbon sources. Since the government buys about half of all cement in the United States, this is a huge step, kickstarting the creation of products for a buyer who is able to eat extra costs. Shifts in public investment are critical but this barely registered as a news item and might be undermined by future administrations.

At a global view, three-quarters of all emissions are from energy use of some kind (electricity, transportation, use of energy in manufacturing, and so on). Global warming is an energy problem and a capital problem. Four-fifths of our energy comes from fossil fuels, and the correlation coefficient for GDP and atmospheric carbon dioxide concentration is 0.96. Before we address those problems, let's cover the physics.

## Earth Energy Imbalance

Begin with the earth energy imbalance. The sun continuously hits Earth with 174,000 terawatts (TW). A terawatt is a trillion watts (W). A watt is unit of power equal to a joule per second (J/s). So to say that the sun hits Earth with so much wattage includes both energy and time that together form force (energy over time). Earth absorbs a massive amount of energy from the sun. For comparison, human civilization runs on a little over 19 TW.

As sunlight strikes Earth, 30 percent bounces back into space due to Earth's albedo effect, the reflectivity due to clouds and ice. So 123,000 TW remain. Consult a Trenberth diagram to see the many vectors of radiation bouncing off clouds or striking the surface.

We measure the energy reaching the planet in watts per square meter (W/m$^2$) of Earth's 510 million square kilometers of surface

---

10  Rebecca Dell, "Rebecca Dell on Decarbonizing Heavy Industry," interview by David Roberts, *Volts*, February 11, 2022, podcasts.apple.com.

area. Averaged over Earth's entire surface area (land and ocean), the solar radiation would deliver 340 W/m². After subtracting for the albedo effect, the power is 240 W/m².

That's the number to memorize: 240 W/m². If incoming energy holds steady, a body must radiate out the exact same amount of energy in order to stay the same temperature. If a body radiates more energy than it receives, the net loss of energy results in cooling. If a body radiates less energy than it receives, the net addition of energy results in warming.

Physicists using the Stefan-Boltzmann law can calculate what Earth's temperature would be based on its size and the energy it receives if it were simply a body radiating energy back to space: -18°C. However, Earth's actual average surface temperature is 15°C, and the 33°C difference is due to greenhouse gasses.[11]

The large majority of the greenhouse "blanket" is water vapor (20°C). We don't control this number. It responds as a fast feedback to air temperature (warmer air holds more moisture), so we influence water vapor only indirectly with warming from other gasses. Carbon dioxide makes up 8.6°C of the blanket. Ozone adds 2.6°C, methane 1.5°C, and nitrous oxide 0.5°C.[12] What we now do is add greenhouse gasses, forcing Earth to absorb extra energy. So far, we've added 1.3°C to the blanket.

The greenhouse effect comes from the change in atmospheric opacity preventing outbound radiation emission. The sun's radiation strikes Earth, and the surface radiates energy back out at infrared wavelengths. Greenhouse gasses make the troposphere more opaque at infrared wavelengths. So radiation rising back to space is blocked. Outgoing radiation increases as the planet warms, but retained energy increases as well. The result is an energy imbalance between energy in and energy out.

What is Earth's energy imbalance? It's monitored through a number of imperfect options including observations of in situ temperature, measurements of ocean surface heat fluxes, satellites gauging thermal expansion of oceans, or combinations of satellite and in situ temperature measurements. I accessed data from NASA's CERES satellite system to generate a running average of top of the atmosphere net flux from March 2000 to November 2022 (latest

---

11 Thomas Murphy, *Energy and Human Ambitions on a Finite Planet* (Oakland: eScholarship University of California, 2021), 11–13.

12 Ibid., 150.

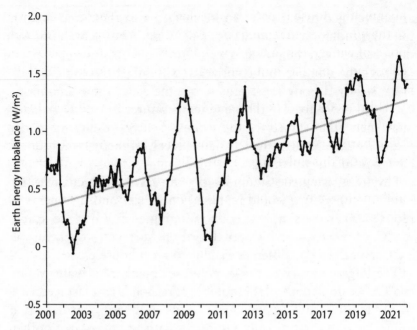

Figure 4.2. Earth Energy Imbalance

available data).[13] As of this writing, the earth energy imbalance approaches 1.3 W/m².

The imbalance is increasing. A joint NASAS-NOAA study published in 2021 found the imbalance had doubled over the prior fourteen years.[14] Until emissions peak, the imbalance will worsen.

Do the math to see that the earth energy imbalance amounts to less than a half percent gap between incoming solar radiation and outgoing infrared radiation. Consider three examples of an Earth energy imbalance.

First example. Small bulbs decorating a Christmas tree are about 0.4 W. Earth's energy imbalance is roughly equal to hanging three small lights over every square meter of the planet's surface. Run them indefinitely. The effect starts small.

Second example. Hang three small Christmas tree lights over every square meter of an average-sized two-hundred-square-meter house (for our purposes, let's round the power to 1.3 W). Then

---

13    EEI data from "CERES_EBAF-TOA_Ed4.2 Subsetting and Browsing," NASA, accessed February 17, 2023, nasa.gov.

14    Norman Loeb et al., "Satellite and Ocean Data Reveal Marked Increase in Earth's Heating Rate," *Geophysical Research Letters* 48, no. 13 (June 15, 2021).

imagine the house is perfectly insulated so any energy introduced to the house-system remains trapped inside. Starting at the freezing point of water, air in a house with Earth's energy imbalance would gain 8.99°C per day and reach water's boiling point in a little over eleven days.[15]

Third example. The Hiroshima atomic bomb released energy equivalent of fifteen kilotons of TNT, or $6.27 \times 10^{13}$ joules of energy. With Earth's energy imbalance the planet gains $5.73 \times 10^{19}$ joules per day. At this imbalance, Earth gains extra energy equivalent to 913,608 Hiroshima atomic bombs every day, or more than 333 million atomic bombs per year.[16] Nine-tenths of the heat goes into the ocean.

## Forcings

Energy imbalance results from radiative forcings. James Hansen defined forcing as "an imposed perturbation (disturbance) of the planet's energy balance."[17] The most important source of forcing is carbon dioxide.

Forcings can be positive or negative. By turning the atmosphere more opaque at infrared wavelengths, greenhouse gasses like carbon dioxide produce a positive forcing by making Earth retain

---

15   A two-square-meter house holds ~2,500 kilograms of air. Air's specific heat capacity is 1,000 joules per kilogram per degree, so the house gains energy at the rate of 260 joules per second (1.3 W × 200 m²/ second), which would raise the temperature 0.0001°C per second ([(1.3 × 200)/1000]/2500) or 8.99°C per day. Or to calculate another way, say we start at 0°C and ask: How long would it take for the house's air temperature to reach 100°C? Divide for 100°C by 8.99°C/day to arrive at 11.13 days. See Murphy, *Energy and Human Ambitions*, 90–3. See also Dawn Stover, "How Many Hiroshimas Does It Take to Describe Climate Change?," *Bulletin of the Atomic Scientists*, September 26, 2013, thebulletin.org.

16   A ton of TNT is $4.18 \times 10^9$ joules, so multiplied by 15 kilotons, the Hiroshima bomb released $6.27 \times 10^{13}$ joules total. But Earth's energy imbalance of 1.3 W/m² multiplied for 510 million square kilometers equals $6.63 \times 10^{14}$ joules per second (1.3 × 1,000,000 × 510,000,000). Multiplied for 60 seconds, 60 minutes, and 24 hours, Earth gains $5.73 \times 10^{19}$ joules per day. Divide this energy gain by the energy released by the Hiroshima bomb. At this imbalance, Earth gains extra energy equivalent to 913,608 Hiroshima atomic bombs every day.

17   Hansen, *Storms of My Grandchildren*, 5.

energy. A negative forcing such as from aerosol pollution reduces the amount of sunlight reaching the surface.

Earth experiences multiple forcing agents simultaneously, so positive forcing is offset by negative forcing. The relationship of greenhouse gas forcings to energy imbalance and rising temperatures is not controversial. It is basic physics.

In addition to positive or negative, forcings can be natural or anthropogenic. Positive forcing emissions are anthropogenic (combusting fuels, changes to land use), but a small amount of carbon dioxide is emmitted from volcanoes and methane, from land. Humans emit sixty times as much carbon dioxide as the annual average from volcanoes, so the latter are small players on human timescales, even if they tip the scales over millions of years. Aerosols scatter sunlight, so they are negative forcers, which can be anthropogenic or natural. Pollution visible to the naked eye is aerosol pollution.

Climatologists use the term "forcing" broadly. Earth's albedo effect from clouds and ice is a natural negative forcing. If we were to seed clouds or brighten them, that would be an anthropogenic negative forcing that augments a natural negative forcing. If we were to plant as many trees as possible to suck up carbon dioxide, branches sticking out of the snow in high latitudes would darken the northern hemisphere and pit carbon dioxide and albedo forcings against each other. In an infamous example of natural negative forcing, Mount Pinatubo's 1991 eruption sent so much sulfur dioxide into the stratosphere that temperatures dropped 0.5°C for a couple of years.

The most obvious natural forcing is variation of solar radiation. The sun's eleven-year cycle of dimming and brightening varies the wattage Earth receives. The IPCC's *First Assessment Report* warned, "The size of the warming over the last century is broadly consistent with the prediction by climate models, but is also of the same magnitude as natural climate variability . . . the observed increase could be largely due to this natural variability."[18] At the time of that report, the earth energy imbalance was likely a bit below 0.75 W/m², so solar forcing was practically indistinguishable from the earth energy imbalance. The imbalance wasn't yet measured precisely by satellites and ocean instruments, so scientists faced a real challenge

---

18 Intergovernmental Panel on Climate Change, *First Assessment Report*, Working Group I, 1990, xii.

to isolate the earth energy imbalance signal from the solar variance noise. With better instruments, this is no longer the problem, but deniers still blame the sun.

Greenhouse gasses make up a vanishingly small part of the atmosphere. Depending on when you read this, carbon dioxide makes up around 420 ppm of air by volume, and it makes up around three-quarters of the greenhouse gas emissions problem but 0.042 percent of the atmosphere. The rest of the anthropogenic greenhouse effect is mostly driven by methane (1,866 parts per billion [ppb] $CH_4$) and nitrous oxide (332 ppb $N_2O$).[19]

It's astonishing to realize the rise from 280 ppm to 420 ppm had such an effect. If we keep going and raise carbon dioxide barely more than one-twentieth of one percent, we could eventually eliminate all ice from the poles.[20]

The IPCC's *Sixth Assessment Report* put effective radiative forcing at 2.72 $W/m^2$.[21] But our aerosol pollution (-1.1 $W/m^2$) masks effects of greenhouse gasses.[22] By the way, this is why we must drop carbon dioxide to 350 ppm rather than the truly pre-industrial 280 ppm. There is so much pollution in the air that reducing carbon dioxide to the pre-industrial level would cool Earth too much.

Positive forcings minus negative forcings equals an Earth energy imbalance of 1.3 $W/m^2$.[23] The IPCC estimated aerosols contributed a counteractive cooling effect somewhere in the range of 0 to 0.8°C. Put differently, pollution masks so much warming that if we cleaned up our air, temperatures could have already risen past 2°C.

---

19   Ibid., TS-35.

20   Instead of using 450 ppm for polar ice elimination, I am using 520 ppm to account for negative forcing from aerosols. The addition of 70 ppm is my crude workaround based on the common target where, accounting for current negative forcings from aerosols, Earth should reach equilibrium at 350 ppm rather than 280 ppm.

21   Intergovernmental Panel on Climate Change, *Sixth Assessment Report*, Working Group I, 2021, 91.

22   Ibid., 69.

23   I want to acknowledge that I am using a different EEI (1.3 $W/m^2$) than found in the IPCC's Assessment Report 6, which puts EEI at 0.79 (0.52 to 1.06) $W/m^2$. IPCC used data through 2018 for its EEI, and here I use much more recent data from NASA's CERES satellite system. This also means the positive and negative anthropogenic effective radiative forcing numbers, even if including natural forcings, add up to a slightly lower number than 1.3. I hope the reader will forgive me for using numbers that don't quite add up, due to their being the most current numbers available.

## Feedbacks

Feedbacks can be amplifiers (warming) or diminishers (cooling) and fast or slow. Water vapor is a fast, amplifying feedback. Clouds are fast, diminishing feedbacks, because they are transient and reflect sunlight. Forcings and feedbacks together convey the basic conceptual framework for climate change.

The first principle of slow feedback is ice sheet size. As ice sheets grow or shrink, the albedo effect changes. Color matters as well. Wet ice is darker, so the albedo effect diminishes if, under warmer conditions, ice is melting and refreezing and melting again. Ice sheets in Greenland and Antarctica are the most significant. The second principle of slow feedback is the change of absorption and release of gasses by the land and oceans. The soil and oceans release more gasses as temperatures increase, the oceans absorb carbon dioxide less effectively over time, and melting tundras will release methane. We can examine the slow feedback mechanism of land and ocean uptake of carbon dioxide by examining the carbon cycle.

## Carbon Cycle

In the same way that carbon dioxide exhaled from animals is inhaled by plants, the planet cycles through natural release and absorption. Humans alter the carbon cycle through deforestation and burning fossil fuels. By 2025, human activity will have emitted 2,600 $GtCO_2$, about half of which will have been emitted in the last three decades.[24] But it doesn't all stay in the atmosphere.

A little under half of carbon dioxide emissions remain in the air (44 percent) while the oceans and land take up the rest (56 percent).[25] Of the latter half, three-quarters go into the ocean, and one-quarter goes into vegetation and soil. When carbon dioxide is taken up by water or land, that is called a sink.

When we emit a tonne of carbon dioxide, sinks absorb one-fifth in the first five years. After a century, 60 percent is absorbed. The

---

24  The number 2,600 is roughly calculated from five years at 41.4 $GtCO_2$ per year added onto 2,390 $GtCO_2$ as of 2020 per Intergovernmental Panel on Climate Change, *Sixth Assessment Report*, Work Group I, 2021, 29.

25  Ibid., 4.

remaining portion lingers far longer. A conventional estimate is that a fifth of the carbon dioxide pulse remains in the atmosphere for as much as half a million years. Without carbon dioxide removal technologies, a fifth of our emissions are effectively permanent on any timescale relevant to human civilization.

Over time, sinks get saturated. Carbon dioxide is less soluble in warm water, so a hot ocean becomes less effective at absorption. Acidic water threatens coral reefs, where a quarter of marine animals live, and harms phytoplankton at the base of the food chain. In higher-emissions scenarios, carbon sinks will take up smaller and smaller percentages of total emissions and leave more in the air (meaning greater permanent damage per tonne emitted).

On longer timescales of tens of thousands to millions of years, atmospheric carbon dioxide drops with rock weathering. As we saw in the previous chapter, rock weathering occurs when rain, acidic with carbon dioxide, chemically reacts with minerals and sediment washes out to sea, where it is buried on the ocean floor. While this process dramatically changes Earth's climate in the long term and is responsible for several mass extinctions, its speed of a little over a tenth of 1 ppm per year is too slow to change climate significantly on timescales meaningful to humans.

## Climate Sensitivity and Why Warming Stops at Net-Zero Emissions

How much will the climate warm before it reaches equilibrium? Climate sensitivity is the response to forcing. The paleontological record shows how temperature responded to past changes in carbon dioxide. Ever since the 1979 Charney report (to which we will return in the following chapter), the canonical number has been 3 °C warming per doubling of carbon dioxide. Math is in the footnote, but doubling carbon dioxide from the pre-industrial 280 ppm to 560 ppm should raise global average surface temperatures around 3 °C.[26]

---

26   IPCC uses coefficient 5.35 such that the logarithmic relationship between forcing and doubling carbon dioxide from 280 to 560 looks as follows: $5.35 \times \ln(560/280) = 3.71$ W/m². To get the temperature corresponding to that forcing, multiply this forcing number by the equilibrium climate sensitivity number 0.75 °C per W/m² (lower range of sensitivity, which might be as high as 1 °C per W/m²; uncertainty remains due to

Equilibrium climate sensitivity designates how much the earth must eventually warm up to rebalance energy radiated with energy received. It's one of several ways to talk about climate sensitivity. Transient climate response and Earth system sensitivity are two other measures for sensitivity of which the reader should be aware.[27] Using equilibrium climate sensitivity, since Earth is out of balance by 1.3 $W/m^2$, the planet must warm up by a certain amount to regain equilibrium between energy absorbed and energy radiated.

Hansen's team examined paleoclimate data to derive a simple climate model from historic deep ocean temperature fluctuations (measured with an oxygen isotope) in relationship to carbon dioxide throughout our Cenozoic Era.[28] Ice cores also provide precise data on levels of carbon dioxide, methane, and nitrous oxide for the last

---

negative forcing from aerosols). See Intergovernmental Panel on Climate Change, *Fourth Assessment Report*, Working Group III, 2007, TS-39. Temperature increase per Intergovernmental Panel on Climate Change, *Sixth Assessment Report*, Working Group I, 2021, 11.

27    By focusing on equilibrium, I've simplified climate sensitivity in this section, but there are a number of different ways to think about sensitivity. The measure of climate sensitivity we have been using here is equilibrium climate sensitivity (ECS). This measure of eventual temperature—the temperature at which Earth regains energy balance—is one of three ways to measure climate sensitivity. A second measure, transient climate response (TCR), gauges temperature at the time carbon dioxide reaches a certain threshold. The 0.45°C rise per 1,000 $GtCO_2$ number from IPCC expresses TCR, and TCR is probably the more common measurement one sees in climate literature. Even if temperatures will eventually rise 3°C for a doubling of carbon dioxide (per ECS), they will not yet have reached that high when the double threshold is first hit. It will take time since heat will not yet have distributed evenly across atmosphere, oceans, and land. Third, Earth system sensitivity (ESS) measures long-term effects of warming, including slow feedbacks such as ice sheet collapse, which by reducing the albedo effect produce additional climate forcing on top of greenhouse gas forcing. ESS indicates doubling carbon dioxide eventually raises temperatures between 1.5°C and 4.5°C. The reader should be aware there are multiple ways to measure sensitivity. See Zeke Hausfather, "Explainer: How Scientists Estimate Climate Sensitivity," Carbon Brief, June 19, 2018, carbonbrief.org.

28    James Hansen et al., "Climate Sensitivity, Sea Level and Atmospheric Carbon Dioxide," *Philosophical Transactions of the Royal Society* 371, no. 20120294 (August 13, 2021).

800,000 years. The team found equilibrium climate sensitivity to be 0.75°C per W/m². (Hansen more recently suggested sensitivity as high as 1°C per W/m².)[29]

Applying this simple formula, current Earth energy imbalance gives us 0.98°C additional warming (1.3 × 0.75). To regain energy-in, energy-out equilibrium, Earth must warm to 2.14°C. That ruins Paris Agreement goals. However, it's not inevitable.

The good news is warming should stop quickly after emissions reach zero. For a long time, many believed a large amount of warming was "in the pipeline," locked in, and inevitable. While it's true constant emissions (constant forcing) means committed warming to reach equilibrium, the problem dissipates quickly after we zero out emissions (reduce forcing). Sea levels will continue to rise as excess energy spreads and melts ice even as carbon dioxide in the atmosphere falls, but ocean and land sinks will draw down atmospheric carbon dioxide. By chance, the warming effect of the ocean and the cooling effect of less atmospheric carbon dioxide cancel each other out.

This good news applies only if emissions drop steeply. Halving emissions would be enough to reduce atmospheric carbon dioxide, but as carbon sinks become saturated, emissions would need to drop closer to zero. An additional threat looms in reduced solar scattering by aerosols, which we must cut, since pollution kills ten million people per year; even if warming is not physically committed, it may be politically committed. But the point is straightforward: positive emissions lead to rising temperature and rising carbon dioxide concentration, zero emissions lead to flat temperature and falling carbon dioxide concentration, and negative emissions lead to falling temperature and falling carbon dioxide concentration. Our best modeling indicates that fifty years after zeroing emissions, we will see a stabilized temperature at ± 0.3°C (the model average is -0.03°C) compared to wherever temperatures are when we reach zero emissions.[30] For better or worse, wherever the temperature is when we zero emissions is where it will stay.

---

29   Hansen et al., "Global Warming in the Pipeline," preprint, July 5, 2023, columbia.edu.

30   Zeke Hausfather, "Explainer: Will Global Warming 'Stop' as Soon as Net-Zero Emissions Are Reached?," Carbon Brief, April 29, 2021, www.carbonbrief.org.

## Temperature Projections and Shared Socioeconomic Pathways

When *Nature* surveyed IPCC authors, more than three-quarters of respondents expected we'd exceed 2°C by century's end. Nearly half the authors expected 3°C. More expected we'd hit 4°C than limit the increase to 1.5°C.[31]

Where one lands on that question implies a political philosophy behind emissions scenarios. The left ought to be much more informed and vocal in the interpretation of emissions scenarios. They all boil down to the well-worn formula about it being easier to imagine the end of the world than the end of capitalism. The IPCC doesn't do political theory. Its projections are thoroughly capitalist futures sprinkled with slight hints of social democracy.

IPCC reports use shared socioeconomic pathways (SSPs), which combine a socioeconomic scenario with a forcing or representative concentration pathway (RCP) number. Five narratives for possible futures were developed representing either high or low challenges to emissions mitigation as well as high or low challenges to adaptation. The five SSP "marker scenarios" each have a baseline scenario where no mitigation of greenhouse gas emissions occurs. The baseline scenario is thus the worst-case projection, given certain inputs or drivers, such as population, urbanization, and GDP. After finding the baseline scenario, which has the highest forcing (RCP4.5 or 6 at the lowest), scientists modeled a couple dozen more scenarios where greenhouse gas forcing is mitigated through several options: policies such as Kyoto-style state-backed limits on emissions, changes to carbon prices, or carbon capture and storage. We will examine in detail the integrated assessment models used for modeling exercises in chapter 8. Adjusting mitigation options gives us a lower RCP number, and it is these lower numbers that tend to show up on IPCC charts repeated in the media (and in this book). For instance, SSP1-1.9 designates socioeconomic pathway 1 plus forcing at 1.9 W/m², but that implies a great deal of exogenous policy mitigating emissions since the SSP1 baseline scenario (that is, without emissions mitigation) has a higher forcing number of RCP4.5 or 6 (depending on when the modeling was done). What

---

31 Jeff Tollefson, "Top Climate Scientists Are Sceptical That Nations Will Rein in Global Warming," *Nature* 599, no. 7883 (November 1, 2021): 22–4.

I'm trying to tell you is this: many of the temperature projections you actually see reported here and elsewhere for various possible futures already assume successful emissions mitigation.

The scenarios point at five possible futures. Not prophecies; SSPs are research tools for examining how a change in one variable might affect others.

Sustainability: low challenges to mitigation, low challenges to adaptation. SSP1 is a sustainable future respecting planetary boundaries and human dignity. We improve health and education opportunities, and we shift to low material growth, reduce resource use, manage the commons, and eliminate inequalities. SSP1 is more or less an outbreak of global social democracy over the next couple decades.

Middle road: medium challenges to mitigation, medium challenges to adaptation. SSP2 extends today's social, economic, and technological trends. Growth is uneven and some states make progress while others fall further behind. Environmental degradation continues while sustainability pledges go unmet. Inequality persists and drives uneven vulnerability to catastrophe. Since SSP2 extends current trends, it is correct to designate this pathway as "business as usual" (not the higher emissions scenario of SSP5).

Regional rivalry: high challenges to mitigation, high challenges to adaptation. SSP3 is the rocky road of regional trade-protectionist nationalism, inequality, and war. Policymakers chase a nightmare of energy security and sideline environmental concerns. States focus on food security within their borders, and public funds drop off for education and technology. Consumption is material-intensive, and inequalities worsen as populations grow in the developing world. As the ecological and social problems stack up, the international community loses interest in addressing the climate crisis. SSP3 seems a likely path and broadly fits the logic of capital, though trade protectionism cuts against growth and may not be tolerated by the ruling class. I expect we will burn through most of our reserves, and in SSP3 (as in SSP5), we burn through them.

Inequality: low challenges to mitigation, high challenges to adaptation. SSP4 is a highly unequal future. Knowledge and capital-intensive sectors benefit while poor societies are relegated to manual labor. Loss of social cohesion leads to conflict. Technological investment in the developed world and diversified fossil fuel and renewable energy systems mean high economic growth unevenly distributed. Ecological concern turns insular to focus on

| Scenario | Near term 2021–2040 | | Mid-term 2041–2060 | | Long term 2081–2100 | |
|---|---|---|---|---|---|---|
| | Best estimate (°C) | Very likely range (°C) | Best estimate (°C) | Very likely range (°C) | Best estimate (°C) | Very likely range (°C) |
| SSP1-1.9 | 1.5 | 1.2 to 1.7 | 1.6 | 1.2 to 2.0 | 1.4 | 1.0 to 1.8 |
| SSP1-2.6 | 1.5 | 1.2 to 1.8 | 1.7 | 1.3 to 2.2 | 1.8 | 1.3 to 2.4 |
| SSP2-4.5 | 1.5 | 1.2 to 1.8 | 2.0 | 1.6 to 2.5 | 2.7 | 2.1 to 3.5 |
| SSP3-7.0 | 1.5 | 1.2 to 1.8 | 2.1 | 1.7 to 2.6 | 3.6 | 2.8 to 4.6 |
| SSP5-8.5 | 1.6 | 1.3 to 1.9 | 2.4 | 1.9 to 3.0 | 4.4 | 3.3 to 5.7 |

Figure 4.3. Temperatures for Shared Socioeconomic Pathways

local issues. This sounds like the second most likely option, not unlike sci-fi dystopias where the rich live in high-tech luxury while the majority suffer in squalor. I don't know why Figures 4.3 and 4.4 from the IPCC didn't include SSP4, but the end-of-century temperature is around 2°C. Oddly, temperatures are so low because, in such an unequal world, a large segment of our population is stuck in hell, unable to develop and so unable to contribute much to emissions. This points to an uncomfortable reality: the worst SSPs from a perspective of human flourishing (SSP3 and 4) are not the worst for temperature (SSP5).

Fossil-fueled development: high challenges to mitigation, low challenges to adaptation. SSP5 sees increasingly competitive market growth. Investments in education and health build the workforce. Capitalism's unabated growth incentive pushes for full exploitation of fossil fuels. The booming economy combined with reduced inequality from human investment leads to better management of ecological problems. The air is cleaned up, the population stabilizes, and geo-engineering controls the heat. It contains all the social and capitalist contradictions expected in a liberal dream scenario.[32]

Of the five main scenarios, two reflect high emissions (SSP3-7.0 and SSP5-8.5). The former is the worst for human suffering, while the latter is the worst-case scenario for warming, and it is often called "business as usual" but would be more accurately described

---

32 Description of SSPs from Keywan Riahi et al., "The Shared Socioeconomic Pathways and Their Energy, Land Use, and Greenhouse Gas Emissions Implications: An Overview," *Global Environmental Change* 42 (2017): 153–68. Figure 4.3 from Intergovernmental Panel on Climate Change, *Sixth Assessment Report*, Working Group I, 2021, 14. Figure 4.4 from ibid., 13.

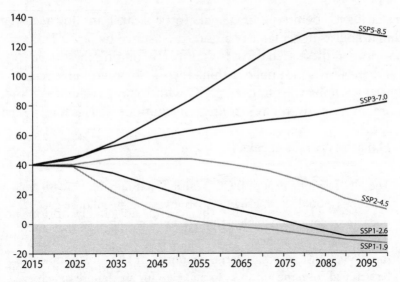

Figure 4.4. Carbon Dioxide (GtCO$_2$/year)

as unabated growth. It's usually what you are reading about if a journalist wishes to sensationalize the worst possibilities. It could still be our future. As discussed above, SSP2 extends current trends and thus is most properly described as business as usual. "Business as usual" is an interesting term for a Marxist, of course, due to the overdetermination of contradictions. Whether we speak of the controversial tendency of the rate of profit to fall, the well-established cycle of crisis, or the rising rate of exploitation, which can be coincident with rising real wages, "as usual," for a Marxist, is anything but a trend line extended indefinitely.

In the middle, SSP2-4.5 is a moderate scenario. It's often treated as where we'll get eventually. Positive forcing currently hovers near 4 W/m$^2$, so it is fair to say we are living in an SSP2-4.5 world now. But current pledges and policies should lead to 2.5–3.2°C by 2100.[33] That means we are on path for a future between SSP3-7.0 and SSP2-4.5 if capitalist governments follow through on commitments. If I had to bet, an SSPx with features modified from SSP2 and SSP3, with pockets of deep inequality from SSP4, would represent the closest to our likely future.

The most optimistic scenarios, SSP1-1.9 and SSP1-2.6, require net-zero emissions by 2050. In my introduction, I implied these

33 "Temperatures," Climate Action Tracker, accessed April 4, 2022, climateactiontracker.org; and Intergovernmental Panel on Climate Change, *Sixth Assessment Report*, Working Group III, 2022, 17.

scenarios exude a type of denial. To be clear, I am not accusing climatologists of denial. My argument, inspired by James Hansen's position on these scenarios, is that merely putting these scenarios out there allows the public to think there is some workable scenario where we limit warming to 1.5°C with today's technology and, more importantly, today's political institutions. This is a mistaken judgment.

Hansen's critique speaks for itself:

> The UN scientific group, IPCC, realized that unfettered fossil fuel emissions would cause growth of atmospheric greenhouse gases to outstrip scenarios in which global warming is limited so as to avoid dangerous consequences. Thus they devised a scenario, RCP2.6 [now SSP1-2.6], in which large quantities of $CO_2$ are assumed to be stripped from the air, so as to make up for any failure to achieve emission reductions.[34]

The rosiest scenario ends in as little as 1°C by century's end, and in the worst scenario, the temperature rises as high as 5.7°C. But the future doesn't stop at the end of this century.

## Effects: Sea Level

Paleoclimate data tells how sea level responds to temperature increases. Carbon dioxide, temperature, and sea level move together; carbon dioxide either drives temperature change (as it did in many of the mass extinctions) or lags behind as a feedback (such as when Milankovitch cycles drop temperature, which lets colder ocean water absorb more carbon dioxide from the air, which drops temperature further). Whatever the order, sea level changes in response. While carbon dioxide levels and temperatures rise and fall nearly simultaneously, sea level lags by one to four centuries.

Multimeter rise may happen linearly with gradually faster speed, or it may be punctuated with smaller, quick jumps (on the order of decades). Surprises happen. While I wrote this book, an Antarctic ice shelf in Wilkes Land, previously thought a stable area, collapsed around the same time we saw simultaneous heat records set in the

---

34  Hansen, "Climate Change in a Nutshell," 51–2.

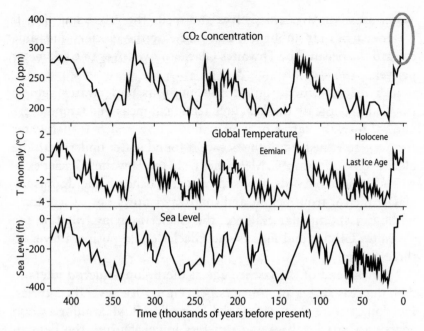

Figure 4.5. Carbon Dioxide, Temperature, and Sea Level

Arctic and Antarctic. What we do know is that, while sea level lags behind, it inexorably follows carbon dioxide and temperature.[35]

An ice sheet is made up of a glacier on land and an ice shelf at the edge floating over the water or resting on land below the waterline. When part of the ice shelf breaks away, it becomes an iceberg and melts. More and more frequently, ice shelves fracture, but the entire ice sheet cannot simply crack and float away since most of the ice sheet is on land. Ice at the continent's edge buttresses the larger volume behind it. Ice buttressing critical West Antarctic glaciers might collapse in a decade's time. If that were to happen, it would hasten the anterior glacier's slide out to sea, but the glaciers wouldn't float away immediately. So much ice will take time.

The "doomsday" Thwaites Glacier and the neighboring Pine Island Glacier in the West Antarctic are held back from the Amundsen Sea by ice at the grounding line that buttresses glaciers against sliding out to sea. The grounding line is the last point at which the glacier sits on solid ground, and it forms the boundary with the floating ice shelf beyond. Extra heat hasn't penetrated the deep ocean yet, but it has warmed the layer at the grounding line's depth.

35   Figure 4.5 adapted from John Englander, "Chart of 420,000 Year History," accessed May 13, 2022, johnenglander.net.

Water seeps underneath shelves and melts the grounding line. If water reaches far enough, it can lubricate the glaciers and slide toward the ocean. The Thwaites Glacier moves over one kilometer per year.

In a recent expedition, an autonomous underwater vehicle traveled underneath the Thwaites Trough. It measured temperature and salinity, the latter of which shows how much is meltwater from a glacier instead of the ocean. It found water underneath the glacier to be 0.8–1.05°C. Meltwater from the Thwaites Glacier exits west.[36] In other words, troughs of water beneath the doomsday glacier melt it from below, and that water flows out to sea. Even without a spectacular collapse, the glacier is dying. Melting the Thwaites Glacier and Pine Island Glacier could push sea level up three meters.

Near the end of the last ice age at the Bølling-Allerød interstadial, the seas rose 4 or 5 meters per century for several centuries. Not long before that rise, at the Last Glacial Maximum 25,000 years ago, sea level was 130 meters lower. During the Eemian Interglacial, around 125,000 years ago, humans lived in a climate a degree warmer than now, but seas were 5 to 10 meters higher. In the Pliocene Epoch, some 3 million years ago, when atmospheric carbon dioxide concentration mirrored today's and temperatures stood a 2.5–4°C above pre-industrial levels, oceans were 15 to 25 meters higher.[37]

Greenland will go first and holds enough ice to raise sea level seven meters. Melting all Antarctic ice would add fifty-eight meters. The loss of all other glaciers would add another third of a meter.[38] Thermal expansion of water will raise levels a bit more.

So far, sea level has risen sixteen centimeters, now nearly four centimeters per decade.[39] For the last fifty years, thermal expansion

36   A. K. Wahlin et al., "Pathways and Modification of Warm Water Flowing Beneath Thwaites Ice Shelf, West Antarctica," *Science Advances* 7.15 (April 9, 2021).

37   Intergovernmental Panel on Climate Change, *Sixth Assessment Report*, Working Group I, 2021, 21; see also Hansen, "Climate Change in a Nutshell," 26

38   Bethan Davies, "Calculating Glacier Ice Volumes and Sea Level Equivalents," Antarctic Glaciers, accessed April 29, 2022, antarcticglaciers .org.

39   Intergovernmental Panel on Climate Change, *Sixth Assessment Report*, Working Group I, 2021, 77.

of water drove half the rise, but now ice melt dominates.[40] Even if emissions zero out, energy spreading throughout the ocean will raise seas for centuries to millennia. If the Western Antarctic Ice Sheet collapses, it won't look like a cataclysmic break in which an iceberg breaks off into the Southern Sea, but cascading collapses of the ice shelves that lock in key glaciers could happen very soon and signal the inevitable changes. In 2016, Hansen's team projected multimeter sea level rise in the next fifty to one hundred fifty years.[41] If all readily available fossil fuels were burned, almost all ice would eventually melt and raise the oceans seventy-five meters.

Even if we decarbonize our economy, sea level will still rise for centuries to millennia as the ocean takes up heat. Depending on the emissions scenario, sea level rises between a half and two meters by 2150. With unabated emissions, sea level might reach seven by 2300. If Antarctic ice cliffs destabilize, sea level could rise by as much as sixteen meters by 2300.[42]

How do temperature targets match up to sea level rise? In the long term of the next two millennia, as seas finally stabilize from what we've done, 1.5°C will raise sea level 2 to 3 meters. If we go to 2°C, 2 to 6 meters. If governments stay on track with current pledges and hit 3°C, 4 to 10 meters. If governments take action but fall short of goals and push up to 4°C, 12 to 16 meters. If nothing is done and we push up to 5°C, north of 20 meters.[43]

Beaches are often a meter or two above sea level at their base. At Eemian-level five- to ten-meter rise, the last time temperatures were this high, much of Bangladesh would be underwater along with large parts of China. Europe would lose the Netherlands and much of its coastlines and cities. The United States would lose much of New York City, Philadelphia, Boston, Washington, DC, much of Florida, and all of New Orleans. And what of the ports? Liberal capitalism is a resilient behemoth, but surely there are limits at the peripheries of permanently stochastic trade disruption and depressions over centuries of sea rise.

---

40  Ibid., 11.

41  See James Hansen et al., "Ice Melt, Sea Level Rise and Superstorm," *Atmospheric Chemistry and Physics* 16 (2016): 3761–812.

42  Intergovernmental Panel on Climate Change, *Sixth Assessment Report*, Working Group I, 2021, 22, 79.

43  Ibid., 79.

## Effects: Isotherms, Wildlife Displacement, and Biodiversity Loss

In *Storms of My Grandchildren*, Hansen recounted a message received after his televised criticism of the Bush administration's censorship of climate science. A viewer from Northeast Arkansas wrote, "I would like to tell you of an observation I have made. It is the armadillo. I had not seen one of those animals my entire life, until the last ten years. I drive the same 40-mile trip on the same road and have slowly watched these critters advance further north every year and they are not stopping." The story struck me, because I grew up in Arkansas and saw armadillos often enough that they were wholly unremarkable. I didn't experience their presence as new. This was a lesson: massive displacement will feel normal for us even as animals push into higher latitudes, if they survive at all.

We talk a great deal about heat, sea level, and mass extinctions, but what of the more mundane? A given temperature line, an isotherm, moves poleward at a rate of sixty kilometers per decade. Species dependent upon specific bands of temperatures will move at a similar speed if they are to survive. In addition to large animals, insects and diseases will move too. A quarter of bee species have disappeared in recent decades, threatening the collapse of human food systems. Three-quarters of bird species are in decline, and a quarter of bats face extinction. Given that birds and bats feed on insects, imagine non-native types of centipedes ballooning in population while we wipe out their predators. Imagine, further, that, just as I in my childhood did not know armadillos were a recent neighbor, future generations experience parasites, malaria, and poisonous spiders as normal.

A sixth mass extinction is a possibility, not a given. The biodiversity crisis sits parallel to and in relation with the broader climate crisis. Animals and plants will adapt in the long term, but that does not mean they can adapt to several degrees over decades. The threats aren't all linear and will surprise us with tipping points. Examining quick changes to ecosystems, the researchers Christopher Trisos et al. projected "future disruption of ecological assemblages as a result of climate change will be abrupt, because within any given ecological assemblage the exposure of most species to climate conditions beyond their realized niche limits occurs almost simultaneously."[44] This study estimated 4°C warming would

---

44   Christopher Trisos, Cory Merow, and Alex Pigot, "The Projected

inflict seven times as many abrupt exposure events to ecological assemblages as compared to only 2°C. The loss of one species in an ecological assemblage affects others. Downstream effects can mean loss of food for humans. This is one reason why carbon dioxide uptake by the ocean is so dangerous. Small fluctuations in ocean acidity damage phytoplankton and threaten the whole food chain.

The extinction rate is already tens to hundreds of times greater than it has been in ten million years. Over the last fifty years, the pace of extinction surpassed anything seen while humans walked the planet.[45] We've so thoroughly changed the makeup of life on Earth that 95 percent of life by weight is humans and our livestock.

## Effects: Heat Waves and Wet Bulb

We already see more intense thunderstorms, tornadoes, heat waves, floods, and wildfires. The IPCC warns of more "compound extreme events" combining multiple hazards (heat waves plus droughts, a flood driven by unusual rainfall, or wildfire from drier conditions). Whatever phenomenal events it exacerbates, simple heat will kill. I suspect heat deaths will increasingly be blamed on victims, as if they did not take proper precautions.

In 2003, a heat wave in Europe killed thirty-five thousand. In 2010, some fifteen thousand Russians perished in another wave. In the final days of editing this book in the summer of 2023, we saw a continuous thirty-six-day stretch surpassing all prior records. In the summer of 2021, a heat dome over the Pacific Northwest killed hundreds of Americans and Canadians—by one count, over a thousand. I was only beginning this book at the time. What I remember is that, when heat records were passed—while people panicked that so few had air conditioning in a region that rarely needs it, while I hoped my friends would survive—our National Public Radio station ran a story on how teenagers on the social media app TikTok were culturally appropriating dances from other

Timing of Abrupt Ecological Disruption from Climate Change," *Nature* 580 (April 8, 2020): 496–501.

45  S. Díaz et al., "Summary for Policymakers of the IPBES Global Assessment Report on Biodiversity and Ecosystem Services," Intergovernmental Science-Policy Platform on Biodiversity and Ecosystem Services, 2019, 12.

teens. At that moment, fatalities still on the rise, I caught a glimpse of a chaotic future where the news is all wrong.

Drier and hotter dry seasons will follow warmer and wetter rainy seasons. A warmer sea surface pairs with more water vapor to result in stronger tropical storms. Los Angeles will adapt to sea level rise, but what if wildfires across the San Gabriel Mountains choke off vectors for fleeing citizens or trucks carrying food for thirteen million people? What of those who die young from constant smoke in the lungs or the elderly with heart conditions? What of cardiovascular or mental health when air quality indices demand we stay indoors for ever-lengthening fire seasons? What of those incarcerated without air conditioning?

"Most jobs are outdoors, either agriculture or construction," Hansen reminded us. "At present, these increasingly difficult working conditions are already beginning to have a measurable effect on economies."[46] David Wallace-Wells delivered a more vivid report from the sugarcane region El Salvador, where "as much as one-fifth of the population . . . has chronic kidney disease, the presumed result of dehydration from working the fields they were able to comfortably harvest as recently as two decades ago."[47] Without dialysis, they can die in weeks. Aridity lines will shift and turn once-fertile land desolate, but the human ability to work lands will hit limits.

Wet bulb temperature is a measure of heat and humidity. At 100 percent humidity, a temperature of 35°C or "wet bulb 35" is lethal to humans within hours. If temperatures are higher, the wet bulb 35 equivalent can be reached at lower humidities. The human body simply is not able to release heat through sweat quickly enough at wet bulb 35, even if you sit still without clothes in the shade. Outdoor work, say a summer harvest, would kill. Wet bulb 26 and 27 are common maximums around the world and are more than enough to kill the vulnerable. My home growing up sometimes reached 45°C or wet bulb 27 while parents sent their kids out to summer football practice. Wet bulb 35 is the nightmare scenario.

Or it was the nightmare until more dire findings indicated wet bulb 30 or 31 is the upper limit. Originally, in 2010, two researchers with an eye toward worst-case scenarios posited wet bulb 35 as the upper limit for a human body (typically within a half degree

---

46   Hansen, "Climate Change in a Nutshell," 34.

47   David Wallace-Wells, *The Uninhabitable Earth* (New York: Tim Duggan Books, 2019), 42.

of 37°C). Skin stays a degree cooler than the internal body, so as to conduct heat. They assumed incorrectly wet bulb "never exceeds 31°C," though their research was otherwise foundational in proposing a limit.[48] One challenge with this estimate is it's difficult to test experimentally, since we are discussing lethal heat conditions. A decade later, a group at Pennsylvania State University tried.

The researchers Daniel Vecellio et al. studied a group of young, healthy individuals to test the wet bulb 35 limit, hypothesizing the real number would be lower. Skin temperature did rise with the ambient thermal environment and sometimes rose above core body temperature, which according to thermodynamics reverses the flow of energy or heat. More importantly, the researchers found a lot of variance in tolerances. The limit is well below wet bulb 35, even though that number is now thoroughly popularized in the lay press. Vecellio et al. concluded two points: there's probably no single wet bulb limit (especially in lower-humidity environments), and the wet bulb 35 limit is too high.[49] If we were to set a provisional limit for what an average young adult in good health can tolerate, it should be more in the range of wet bulb 30 or 31.

Recent data from the tropics indicate wet bulb 35 events crop up from time to time, if only briefly for an hour or two. Collecting data from weather stations, the researchers Colin Raymond et al. found many instances of wet bulb 31 and 33, and two stations reported daily maximum values above wet bulb 35.[50]

In all this talk of upper wet bulb limits, we haven't yet accounted for future warming. Vecellio et al. warned, "Results suggest that, under the business-as-usual . . . emissions scenario, [wet bulb temperature] could regularly exceed 35°C in parts of South Asia and the Middle East by the third quarter of the twenty-first century."[51] Remember that thousands died in heat waves in Europe in 2003

---

48   See Steven Sherwood and Matthew Huber, "An Adaptability Limit to Climate Change Due to Heat Stress," *PNAS* 107, no. 21 (May 3, 2010): 9552–5.

49   Daniel Vecellio et al., "Evaluating the 35°C Wet-Bulb Temperature Adaptability Threshold for Young, Healthy Subjects," *Journal of Applied Physiology* 132, no. 2 (January 28, 2022): 340–5.

50   Colin Raymond, Tom Matthews, and Radley M. Horton, "The Emergence of Heat and Humidity Too Severe for Human Tolerance," *Science Advances* 6, no. 19 (n.d.): eaaw1838.

51   Vecellio et al., "Evaluating the 35°C Wet-Bulb Temperature Adaptability."

and Russia in 2010. Those events never surpassed wet bulb 28, which is more than lethal enough for the elderly, for children, and for those with health conditions.[52]

## Effects: Crops

Famine will hit more frequently as well. Crop yields will generally decline as the world warms, and there will be more frequent production shocks. By volume, corn is the most grown crop in the world. The researcher Michelle Tigchelaar found that in the four countries producing two-thirds of the world's corn—the United States, China, Brazil, and Argentina—production of corn would decline 8–18 percent at 2°C, or 19–46 percent at 4°C.

We can project crop losses in part because we study how crops have reacted to warming thus far. Tropical fruits will likely do fine with a little heat, but the staples on which we have survived for a hundred centuries suffer. As conservative an organization as the World Bank's Potsdam Institute projected crop yields will drop 30 percent at 1.5–2°C and nearly 60 percent at 3–4°C.[53] In the most optimistic numbers I've seen, depending on the crop, yields drop 3.1–7.4 percent per degree of warming.[54] Yield decline will require us to use more land and water in a struggle to maintain subsistence levels. But we will hit unforeseen tipping points.

For developed nations, famines might be annoyances negotiated by purchasing food at steep prices in hard times. Perhaps the Global North will complain of inflation at the grocery store and nervously worry about central banks toying with interest rates, while elsewhere people starve. If supply chains fail and shelves go empty, maybe the comfortable in the North can joke about finding a diet that works. Outside the walls, the world's poor won't have the caloric leeway to laugh.

Several billion people already spend most of their income on food. What happens when economic lag joins dead crops, ecosystem

---

52  Andrea Thompson, "Heat and Humidity Are Already Reaching the Limits of Human Tolerance," *Scientific American*, May 8, 2020, scientific american.com.

53  Hans Joachim Schellnhuber et al., "Turn Down the Heat," Potsdam Institute for Climate Impact Research and Climate Analytics, 2014.

54  Adam Aton, "For Crop Harvests, Every Degree of Warming Counts," *Scientific American*, August 16, 2017, scientificamerican.com.

disruption, heat stress, and social disintegration? It's easy to survive wet bulb events with air conditioning. But many people in the world have no air conditioning, and many who think they do will face blackout conditions when stressed grids face impossible demand. I live in the richest nation in the history of the world, but that we cannot seem to tax ourselves to update crumbling infrastructure doesn't bode well for adaptation. Many do not live near rising oceans but do live near trees or windy plains primed for wildfires. Vast regions may need evacuation.

# 5

# What They Knew and When They Knew It

*In various places people are surprised. What's eating them, these students, the little dears, our favorites, the darlings of civilization? What's up with them? Those who are saying this are playing the fool, this is what they are paid to do.*

—Jacques Lacan

*The utopia of togetherness is a lie. Environmental justice means acknowledging that there is no whole earth, no "we," without a "them." That we are not all in this together.*

—China Miéville

"Science was of no country and of no sex," begins Smithsonian Institute Professor Joseph Henry before the assembled gentlemen of his field. It is August 23, 1856, and over the next few minutes, he will reveal the effect of carbon dioxide. "The sphere of woman embraces not only the beautiful and the useful, but the true."

The paper he reads, "Circumstances Affecting the Heat of the Sun's Rays," is the work of a woman named Eunice Foote.[1] She is a women's rights activist from upstate New York, an original signatory to the Seneca Falls Declaration of Sentiments, and an amateur scientist. For unclear reasons, though almost certainly indicated by Henry's awkward reminder that women sometimes say true things, Foote's work is presented by a man before the conference. What she's discovered is remarkable. In a page-and-a-half-long paper

---

1 Eunice Foote, "Circumstances Affecting the Heat of the Sun's Rays," *American Journal of Science and Arts* 22 (1856).

cramming in three graphs of data, she becomes the first person to say in print that if carbon dioxide levels rise, the planet warms.

She starts, "My investigations have had for their object to determine the different circumstances that affect the thermal action of the rays of light that proceed from the sun. Several results have been obtained." Several results indeed.

Placing thermometers inside empty cylinders, Foote finds, "The action of the sun's rays was found to be greater in moist than in dry air." Next, making history with the archaic term for carbon dioxide, Foote explains: "The highest effect of the sun's rays I have found to be in carbonic acid gas." She finishes with an offhand comment predicting what the paleoclimate record will eventually show. "An atmosphere of that gas would give to our earth a high temperature; and if as some suppose, at one period of its history the air had mixed with it a larger proportion than at present, an increased temperature from its own action as well as from increased weight must have necessarily resulted."

## Discovering the Vast Experiment

Today, we often hear how long the fossil fuel companies knew. For nearly a half century before Greta Thunberg first sat outside the Swedish Parliament one August day in 2018, Exxon and the rest knew. But in fact, the history stretches further. The basics discovered by Eunice Foote and John Tyndall were understood before the American Civil War.

As early as the 1820s, the French physicist Joseph Fourier discovered Earth should be colder than it is, and without using the term "greenhouse effect," he essentially described it by attributing anomalous warmth to air. Then, at the Royal Institution of Great Britain, on June 10, 1859, the Irish physicist John Tyndall presented research on what we'd now call infrared absorption of various gasses. What Foote had not understood—the role of infrared absorption in the greenhouse effect—Tyndall rigorously tested. Several times over the course of his three-page paper, he credited studies by Claude Pouillet and William Hopkins as the link between Fourier and himself, so whether Foote influenced Tyndall's work is unclear. What Tyndall isolated was the different effects of different gasses. Water vapor was a strong greenhouse gas, but so was "dry coal-gas" or carbon dioxide. "When heat is absorbed by the planet, it is so changed in quality that the rays emanating from the planet cannot get with the same freedom back into space," Tyndall explained of what we now call Earth energy imbalance. "Thus the atmosphere admits of the entrance of solar heat, but checks its exit; and the result is a tendency to accumulate heat at the surface of the planet."[2]

Tyndall expanded on his findings in his 1863 book *Heat Considered as a Cause of Motion*. In a few brief sections on hydrocarbon experiments, he described how olefiant gas permits entrance of sunlight but blocks the exit of heat: "Under such a canopy, trifling as it may appear, and perfectly transparent to the eye, the earth's surface would be maintained at a stifling temperature."[3] He speculated that distant planets could be warmed by modifying their "atmospheric envelope."

Oddly enough, Friedrich Engels and Karl Marx closely followed Tyndall's research during this time, which coincided with the 1863–67 period when Marx finished *Capital*. Tyndall was one of several scientists providing their inspiration on metabolism and agriculture, but not climatology. Engels read Tyndall's book in 1866 and notified Marx. Though Marx likely read it immediately, perhaps through a library in London, we only have proof in an 1870 edition of *Heat* heavily marked up with Marx's handwriting.[4] Unbearably ironic is Marx's missed opportunity, had he seen the implications, to incorporate those findings into his theory of surplus value, the law of capitalist accumulation, and the movement of capital as an independent social force in *Capital*.

In the 1890s, the Swedish chemist Svante Arrhenius took an interest in what fluctuating carbon dioxide might do to Sweden. His calculations implied a 4–5°C drop for halving carbon dioxide and a 5–6°C increase for doubling (remarkably close to the 3°C number on which climatology would eventually settle). Arrhenius

---

2   John Tyndall, "On the Transmission of Heat of Different Qualities through Gases of Different Kinds," *Notices of the Proceedings at the Meetings of the Members of the Royal Institution of Great Britain* 3 (1862): 155–8, 158.

3   John Tyndall, *Heat Considered as a Mode of Motion* (New York: D. Appleton, 1869), 451.

4   See Pradip Baksi, "MEGA IV/31: Natural-Science Notes of Marx and Engels, 1877–1883," *Nature, Society, and Thought* 14, no. 4 (October 2001), 385. My thanks to John Bellamy Foster's correspondence for helping me explore this question on Marx's knowledge of Tyndall.

and his colleagues imagined how nice it would be if cold northern Europe were warmed by carbon dioxide (from volcanoes, that is, until a colleague of Arrhenius asked what would happen if we kept combusting coal). In between his climate calculations, he won the Nobel Prize in Chemistry in 1903 for his electrolytic dissociation theory, but he circled back to what coal injection might accomplish in his 1908 book *Worlds in the Making*.

"Remarkable Weather of 1911," read a headline by Francis Molena in the magazine *Popular Mechanics*. Just a few years after Arrhenius suggested coal's role, Molena ran the headline: "The Effect of the Combustion of Coal on the Climate—What Scientists Predict for the Future." The 1912 article worked through various weather anomalies from the year before and explained the greenhouse effect, asking readers whether the two might be related.

In 1938, before the Royal Meteorological Society, the British inventor and amateur climatologist Guy Callendar presented "The Artificial Production of Carbon Dioxide and Its Influence on Temperature." He calculated human activity had injected 150 $GtCO_2$ into the atmosphere, and he estimated how this overwhelmed natural sinks and changed temperature. In her history of these discoveries, to which I am indebted, Alice Bell contextualized the typical reaction in Callendar's time: "He didn't necessarily want to ring any alarm bells. Fossil fuels give us heat and power, after all, and he thought global warming could be useful, too—hold off another ice age."[5]

Soon, climate change crossed the Cold War. The USSR began monitoring the Arctic in the twenties and built research stations and ice breaker ships with an eye to improved fisheries and transport. In the 1930s, Greenland had been a strategic stopover for flights to Europe, and after Germany occupied Denmark, the latter authorized US military bases on the island. By 1947, the Swedish glaciologist Hans Ahlmann confirmed the poles were rapidly warming and, in a secret Pentagon meeting, told military brass how the changes would free up Soviet bases in the north. The Pentagon ramped up research on Arctic weather and the cryosphere.[6]

---

5  Alice Bell, "Climate Change History," *Slate*, December 20, 2021, slate.com. See also Alice Bell, *Our Biggest Experiment* (Berkeley, California: Counterpoint, 2021).

6  Catherine Jex, "Climate Change Research Was Born in the Cold War," Science Nordic, January 21, 2017, sciencenordic.com.

In 1950, *Time* published research from the American Meteorological Society claiming the world was hotter. The article, hidden in the back of a science section, evaded public attention. Within a few months, *Scientific American* ran an article on "The Changing Climate," and the *New York Times Magazine* covered the phenomenon three years later.

The two missing puzzle pieces were (1) to what extent the changes were attributable to human activity, and (2) measurements for carbon dioxide concentration. Nobody yet had a good baseline. Carbon dioxide measurements varied widely. Into this vacuum stepped the Caltech postdoc Charles David Keeling. Before long, his eponymous Keeling Curve showed a severe problem.

In 1957, the US Weather Bureau planned to spend a year finding a definitive carbon dioxide level to establish a baseline. The bureau wanted Keeling to take measurements with an infrared gas analyzer in Antarctica and Hawaii, and the oceanographer Roger Revelle recruited Keeling to base his research for the bureau at the Scripps Institution of Oceanography in La Jolla, California, where carbon dioxide measurements are disseminated to this day. On the north slope of the Mauna Loa volcano, Keeling took readings from the pristine Pacific air. On March 29, 1958, the carbon dioxide concentration was 313 ppm, or thirty-three parts above pre-industrial levels.

For two months, concentrations rose impossibly fast. Power glitches interrupted measurements between May and July, but when they resumed, concentrations had dropped. Next spring, carbon

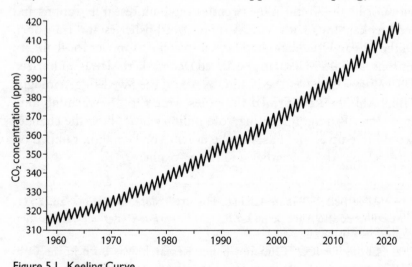

Figure 5.1. Keeling Curve

dioxide rose again. Keeling later recalled "I became anxious that the concentration was going to be hopelessly erratic." After a year, the erratic pattern made perfect sense. In the spring and summer, vegetation growth in the northern hemisphere draws down carbon dioxide to a low point by September or October. When trees lose leaves in the fall and winter, they give it up and, by May, carbon dioxide reaches its apex.

The year of measurements turned into permanent data collection at the Mauna Loa station, producing a pattern that zig-zags about 7 ppm in the course of a year but which climbs more steeply with each passing decade (currently rising about 2.5 ppm per year). With these data in hand, Keeling calculated fossil fuel emissions and compared them against actual atmospheric concentrations to see that the land and oceans were absorbing only half the carbon dioxide. The rest was building up in the air. We are now north of 420 ppm.[7]

In November 1965 the president's Science Advisory Committee transmitted a report on carbon dioxide to Lyndon B. Johnson. It explained rock weathering and the overwhelming injection of carbon dioxide from fossil fuels, which the report estimated outpaced weathering a hundred times over (the real pace is far worse). Authors credited Arrhenius and detailed Keeling's measurements. They didn't hedge: "We can conclude with fair assurance that at the present time, fossil fuels are the only source of $CO_2$ being added to the ocean-atmosphere-biosphere system."[8]

The twenty-three-page appendix on atmospheric carbon dioxide is painful to read now if your working assumption was that climate change is a recent discovery. It was 1965 and the authors estimated it would take 150 years to burn through all fossil fuel reserves, a remarkably close figure given the limits of available evidence. Projecting a 25 percent rise in carbon dioxide concentration by the turn of the millennium, the panel estimated temperatures would rise between 0.6°C and 4°C and finished ominously: "Throughout

---

7 Chart from "Mauna Loa Record—Color Graphic," Scripps CO2 Program, accessed May 13, 2022, scrippsco2.ucsd.edu. See also "Keeling Curve," American Chemical Society, accessed April 5, 2022, acs.org; "Charles David Keeling Biography," Scripps $CO_2$ Program, accessed April 5, 2022, scrippsco2.ucsd.edu; and "The Early Keeling Curve," Scripps $CO_2$ Program, accessed April 5, 2022, scrippsco2.ucsd.edu.

8 "Restoring the Quality of Our Environment," Environmental Pollution Panel of the President's Science Advisory Committee, November 1965, 119.

his worldwide industrial civilization, Man [*sic*] is unwittingly conducting a vast geophysical experiment."[9]

## What the Oil Men Knew

A symposium convened in 1959 to commemorate a century since the first commercial oil well in Pennsylvania. The physicist Edward Teller cautioned industry men and officials against submerging "all the coastal cities." In 1965, the president of the industry front group American Petroleum Institute received a scary report and told colleagues, "There is still time to save the world's peoples from catastrophic consequences of pollution, but time is running out . . . by the year 2000 the heat balance will be so modified as possibly to cause marked changes in the climate beyond local or even national efforts."[10]

The Exxon scientist James Black wrote to the vice president of Exxon Research and Engineering in summer 1978. He attached a presentation he'd given for the company and reminded his boss it was the same information he showed a year prior. The talk was called "The Greenhouse Effect."

From Keeling's measurements, Black said, "current scientific opinion overwhelmingly favors attributing atmospheric carbon dioxide increase to fossil fuel combustion."[11] Black covered Earth energy imbalance and crudely guessed doubling carbon dioxide would raise temperatures up to 3°C overall and 10°C at the poles. The conclusion set a maddeningly optimistic timeframe: "Present thinking holds that man has a time window of five to ten years before the need for hard decisions regarding changes in energy strategies might become critical."[12]

Sleuthing through these documents is as exhilarating as it is enraging. Exxon's internal data accumulated. By 1979, an internal report affirmed that "the present trend of fossil fuel consumption will cause dramatic environmental effects before the year 2050."[13]

---

9   Ibid., 126.

10   As quoted in Malm et al., *White Skin, Black Fuel*, 24–5.

11   James Black, "The Greenhouse Effect," Exxon Research and Engineering Co., 1978, 4.

12   Ibid., 2.

13   As quoted in Geoffrey Supran and Naomi Oreskes, "Assessing ExxonMobil's Climate Change Communications (1977–2014)," *Environmental Research Letters* 12, no. 8 (August 23, 2017): 084019.

A "$CO_2$ Greenhouse Communications Plan" took shape in 1980. The goal was "improved recognition of Exxon as a center of scientific and technological excellence."[14] Company tankers and drilling rigs would serve as laboratories. Through media briefs, college campus organizing, think tanks, and ads, the company would seek to shape public conversation "since future public decisions aimed at controlling the build-up of atmospheric $CO_2$ could impose limits on fossil fuel combustion."

In the fall of 1980, company scientist Henry Shaw told the National Commission on Air Quality about Exxon's knowledge. Agriculture, ecosystems, and sea level would be affected. The changes "can occur within a century, rather than over millennia," and "increases are expected to persist for hundreds of years."[15] Governments must shoulder costs, since the changes "will be adverse to the stability of human and natural communities." The research was buried.

## The Myth of the Global Cooling Consensus

By the nineties and aughts, it was canon for conservatives that scientists once agreed we were headed for a new ice age. Heat concerns, the false history said, are just as misguided. In a comprehensive survey of the cooling myth in peer-reviewed literate and popular media, Peterson et al.'s paper "The Myth of the 1970s Global Cooling Scientific Consensus" showed that, while the infant field of climatology supported global warming, media sensationalism and rightful concern over aerosols drove a misunderstanding that persists today.[16] No such consensus about cooling ever existed.

Nevertheless, you find any number of people who will post magazine covers from the seventies as a counterweight to an IPCC report today. Some even claim they personally remember "when the

---

14  N. R. Werthamer, "$CO_2$ Greenhouse Communications Plan," Exxon, July 8, 1980.

15  Henry Shaw, "National Commission on Air Quality $CO_2$ Workshop Draft Statement of Findings and Recommendations," Exxon Research and Engineering Co., December 5, 1980.

16  Thomas Peterson, William Connolley, and John Fleck, "The Myth of the 1970s Global Cooling Scientific Consensus," *Bulletin of the American Meteorological Society* 89, no. 9 (September 1, 2008): 1325–38.

scientist feared global cooling" or something to that effect. Let us consider the origins of this collective memory flaw.

Temperature records started in the 1870s, and a long-term analysis published by J. Murray Mitchell in 1963 seemed to show Earth in a cooling phase for a couple of decades. Subsequent investigation showed this was a regional trend confined to the Northern Hemisphere, while the Southern Hemisphere warmed. Contemporaneously, we'd learned more about ice ages. We'd known since the nineteenth century that Earth was covered in ice in the past but figured these were rare periods. New paleoclimate data in the late sixties and early seventies revealed, on the contrary, the climate was highly variable with short interglacials sprinkled among long ice ages.

The geophysicist Milutin Milankovitch discovered as early as 1930 a shift in sunlight due to changes in Earth's tilt on its axis as well as the oscillation in orbit from circular to elliptical. With a new understanding of a variable climate in the seventies, Milankovitch's ignored calculations suddenly found an audience. Add in discoveries about the cooling effect of aerosols, and we see the setup for a question about cooling. If temperature changed over time, if ice ages were regular events, if Earth's tilt and orbit changed predictably and drove ice ages, and if aerosols cool, then a calamitous ice age might be our fate. At the "Symposium on Global Effects of Environmental Pollution" in 1968, Reid Bryson argued human pollution drove cooling.

In 1971, S. Ichtiaque Rasool and Stephen Schneider wrote a paper titled "Atmospheric Carbon Dioxide and Aerosols: Effects of Large Increases on Global Climate," which would later turn into a bit of a foundational document for the myth of a global cooling consensus. Unfortunate, because Rasool and Schneider did good work extending nascent climate models to the problem of aerosols even if they overestimated the effect. But it was also the same decade that key players in climatology (Jule Charney, James Hansen, Syukuro Manabe, and Richard Wetherald, to name a few) knew the planet was warming.

In 1973, *Science Digest* published "Brace Yourself for Another Ice Age." *Science News* printed "Climate Change: Chilling Possibilities." The *New York Times* ran articles in both directions: "Scientists Ask Why World Climate Is Changing; Major Cooling May Be Ahead" (1975) and "Warming Trend Seen in Climate; Two Articles Counter View that Cold Period Is Due" (1975). Republican senator James Inhofe likes reminding people of a 1975 *Newsweek* article titled "The Cooling World." *Time* said "climatological Cassandras" prophesied cooler days. No doubt the media dumped the chilling fear on readers in the seventies. Popular books said the world was getting colder. One predicted billions of deaths by 2050. The evidence for global cooling was thin to nonexistent, so what happened? Peterson's team suggested the mix-up came from the relentless drive for eyeballs and the easy hook of a prior catastrophe narrative.

Survey the literature to see what seventies climatologists thought. From 1965 to 1979, seven peer-reviewed articles indicated cooling while forty-four indicated warming. Cooling articles were cited infrequently, and as Peterson et al.'s survey noted, "only two of the articles would, according to the current state of climate science, be considered 'wrong' in the sense of getting the wrong sign of the response to the forcing they considered."[17] In the late sixties, the literature is split between warming and cooling, with very few papers on either, but by the early seventies warm conclusions outpaced cool. Those two wrong articles were split between cool and warm. In short, comparatively few scientists expected cooling. Those who did overestimated aerosols, which is partly attributable to unpredictability in how governments would regulate pollution.

Why does none of this history seem to matter for those who pedantically, maybe with a hint of contempt, smile and remind us that scientists used to believe the planet was cooling? This denial we must keep in mind. What the myth of the global cooling consensus shows is that people are able to construct completely imaginary pasts to justify emissions today. Against all evidence, people will insist they vividly remember something that didn't happen.

## The Charney Report and Exxon's Response

Drought and famine drew attention to vulnerabilities. In 1977, an independent coterie of scientists called the Jasons told the US Department of Energy doubling carbon dioxide would boost temperatures 2.4°C. The Carter administration took notice.

---

17 Peterson et al., "The Myth of the 1970s Global Cooling Scientific Consensus," 1331.

The National Academy of Sciences charged the meteorologist Jule Charney to research the matter. His team included a young climate modeler from the NASA Goddard Institured for Space Studies, James Hansen. The pivotal 1979 Charney Report concluded doubling carbon dioxide would raise temperatures 3°C ± 1.5.

The Charney Report explained feedbacks but said changes were decades away. Decadal timescales are no match for shortsighted bourgeois policymakers. One Jason's allegory: "When you go to Washington and tell them that the $CO_2$ will double in 50 years and will have major impacts on the planet, what do they say? . . . come back in forty-nine years."[18]

In the 1980s, oil men and the broader scientific community worked in parallel. Shaw and Exxon colleague P. P. McCall cited Charney's report when in 1980 they described flooding the capital: "If the Antartic [sic] ice sheet which is anchored on land, should melt, then this could cause a rise in the sea level . . . Such a rise would cause flooding in much of the U.S. East Coast including the state of Florida and Washington D.C."[19]

Scientists with Exxon Research and Engineering Company figured governments must intervene. ER&E launched multiple major research projects for several years starting in 1979. At a cost of a million dollars a year, Exxon outfitted a supertanker with instruments to measure carbon dioxide in the southern Atlantic, the Gulf of Mexico, and the Indian Ocean. Former Exxon scientist Edward Garvey, who worked on the ship, described the sheer volume of information generated as a "data monster." They discovered clear signals. "We were doing some serious science," Garvey said. "We were generating what we thought was state of the art information. We were doing science that we didn't think in any way, shape, or form would be questioned . . . The question was how fast, how much, and what kind of impacts would it have?"[20] They told management what they found.

Management cut budgets and crushed the research in 1982.

---

18   As quoted in ibid., 173–4.

19   Henry Shaw and P. P. McCall to T.K Keto, "$CO_2$ Greenhouse Effect," Exxon Research and Engineering Co., December 18, 1980, 3.

20   Jason Breslow, "Investigation Finds Exxon Ignored Its Own Early Climate Change Warnings," PBS Frontline, September 16, 2015. pbs.org.

## Merchants of Doubt

The US National Academy of Sciences tapped the reactionary physicist and alum of the Manhattan project Bill Nierenberg to assess the state of knowledge. Nierenberg was the first of four key physicists who laid the groundwork for reactionary climate denial. The others, including Fred Seitz, Robert Jastrow, and Fred Singer, were all involved in the tobacco lobby, the chlorofluorocarbon lobby, or Reagan's Strategic Defense Initiative, also known as Star Wars. They are the group that science historians Erik Conway and Naomi Oreskes called "merchants of doubt." All reappear in support of civilization's most existential threat as the founding fathers of a capitalist denialist ISA.

In three of the assessment's five chapters, scientists explained the changes in the pipeline. However, in the two chapters written by economists and consultants, there was "enormous uncertainty" around future emissions, which would be decades off anyway. Nierenberg backed up the rosier view. If people migrated en masse, chased off by collapsing ecosystems, he wrote, "It is extraordinary how adaptable people can be."[21]

In an uncertain environment, reactionaries exploited caution. The Environmental Protection Agency started reporting on greenhouse gasses, and Nierenberg's report handed Reagan an out. The *New York Times* wrote, "The Academy found that since there is no politically or economically realistic way of heading of the greenhouse effect, strategies must be prepared to adapt to a 'high temperature world.'" But they hadn't found this. Economists simply said it. Nierenberg backed them up.[22] It was not "politically or economically realistic" to save the planet.

On a sweltering June day in 1988, NASA scientist James Hansen appeared for Senate testimony. He gave three emissions scenarios and charted their implications. By the way, the path we took, Scenario B, predicted temperatures we see today with eerily near-perfect accuracy. Hansen broke down a choice between rapid decarbonization and unabated growth. "Global warming has reached a level

---

21  Naomi Oreskes and Erik M. Conway, *Merchants of Doubt* (New York: Bloomsbury Press, 2010)., 181.

22  Contra the *Times*, Oreskes and Conway noted, "The Academy hadn't *found* that; the committee had *asserted* it. And it wasn't the Academy; it was Bill Nierenberg and a handful of economists." Ibid., 182.

such that we can ascribe with a high degree of confidence a cause and effect relationship between the greenhouse effect and observed warming," said Hansen. The public was alerted. The *New York Times* quoted his warming of global warming in progress. "It is already happening now."[23]

George H. W. Bush got clever on the 1988 campaign trail: "Those who think we are powerless to do anything about the greenhouse effect forget about the 'White House effect.'" The merchants of doubt redoubled efforts and put their credentials to use. Robert Jastrow was founder and prior director of the NASA Goddard Institute for Space Studies. Fred Seitz was a former president of the US National Academy of Sciences who cashed in by sowing confusion about cigarettes and lung cancer for Big Tobacco. Jastrow, Seitz, and Nierenberg established the George C. Marshall Institute in 1984 to promote Reagan's Strategic Defense Initiative and the war profiteering industry, but the end of the Cold War threatened its raison d'être. No matter. A new target emerged for the conservative think tank.

At the Marshall Institute, the three misrepresented Hansen's data and named a new culprit: solar variation. Their paper circulated widely around the White House. Bush sent his secretary of state to the first IPCC meeting, sought funds for climate research, and intended to join the UN's climate convention. But conservative think tanks lobbied aggressively, and by 1991, Jastrow bragged to the American Petroleum Institute, "It is generally considered in the scientific community that the Marshall report was responsible for the Administration's opposition to carbon taxes and restrictions on fossil fuel consumption."[24]

Meanwhile, Fred Singer continued to cast doubt by undermining Roger Revelle, the Scripps Institute director who at Harvard taught a rising star in green politics: Al Gore. Singer was an experienced merchant of doubt, having denied the links between cigarettes and cancer, the cause of acid rain, and the danger that CFCs posed to the ozone. Singer convinced Revelle to coauthor a paper neutrally named "What to Do About Greenhouse Warming: Look Before You Leap." When Singer downplayed effects and Revelle corrected the numbers, Singer reinserted placid conclusions. It's

23   Philip Shabecoff, "Global Warming Has Begun, Expert Tells Senate," *New York Times*, June 24, 1988, nytimes.com.

24   Oreskes and Conway, *Merchants of Doubt*, 190.

unclear whether Revelle saw the final paper before publication, but he couldn't correct its distortions. He died of a heart attack only months later. Did Revelle write the conclusion, "*The scientific base for a greenhouse warming is too uncertain to justify drastic action at this time*"? Not according to Revelle's student Julian Lancaster, who argued his mentor was tricked. Lancaster believed this was done to harm Gore's agenda. But with Revelle's name on the paper, commentators eventually attacked Al Gore's claims in his 1992 book *Earth in the Balance* with the supposed words of his old professor.

As agents of the capitalist denialist ISA, the merchants of doubt pioneered a menu of tactics: edit selectively, cherry-pick data, blame the sun, fret over costs, personally attack climatologists, and most of all, downplay catastrophes as manageable. By the 1990s, climatology was polarized in the culture war. Nierenberg argued warming would reach no further than a degree by the end of the twenty-first century, a mark passed by the century's second decade.

## The IPCC and the UNFCCC

Just when Hansen's now-legendary Senate testimony alerted the public, the UN established the Intergovernmental Panel on Climate Change in 1988. The organization writes large assessment reports every five to seven years and smaller, topical reports in the interim. They assess the literature in three parts. Working Group I assesses physical sciences. Working Group II assesses impacts and vulnerability. Working Group III assesses mitigation. The IPCC serves as a clearinghouse for climate research.

The final step before publication is the Summary for Policymakers, a much shorter document only a few dozen pages long. If you are not a subject matter expert, chances are any quote you have ever heard from an IPCC report was from a Summary for Policymakers. Representatives from member nations pore over summaries line by line and often dispute findings in accord with their respective states' interests. Member governments can object to wording, but author scientists write the text. The process turns contentious.

More than three hundred scientists worked on the *First Assessment Report* in 1991. A year later, at the Earth Summit in Rio, 154 nations (now 197) signed the UN Framework Convention on Climate Change (UNFCCC). Signatories committed to prevent "dangerous anthropogenic interference in the climate system." It set no binding limits.

With the *Second Assessment Report* in 1995, tensions flared. The atmosphere was warming, but this alone didn't establish the cause. The modeler Benjamin Santer was charged with describing causation, but that autumn someone leaked his draft. Congressional Republicans invited testimony from a coal industry consultant. Democrats responded with their own expert from NOAA, who tried to explain how the model in question deliberately screened out irregularities such volcanic eruptions in order to predict accurately. The details drew little public attention. "I doubt Congress will do anything foolish," Bill Nierenberg told Fred Seitz. "I can also tell you that at least one high-level corporate advisor is advising boards that the issue is politically dead."[25]

A parallel fight erupted over the Summary for Policymakers when American, Saudi, and Kuwaiti delegates opposed the attribution to human causes in Santer's leaked draft. IPCC cochair Sir John Houghton assigned a drafting group the role of hashing out a key sentence: "The balance of evidence suggests that there is a [blank] human influence on global climate."[26] Santer wanted the word "appreciable." After objections, the group settled on "discernible": "The balance of evidence suggests that there is a discernible human influence on global climate."[27] In the *Wall Street Journal*, Seitz attacked the IPCC as rife with "disturbing corruption" and mercilessly slandered Santer's character.

## Oil on the Road to Kyoto

In December 1997, signatories of the UNFCCC met in Kyoto for the third annual Conference of Parties (COP 3). UN secretary-general Kofi Annan announced, "The risks of climate change pose the most critical and pervasive environmental threats ever to the security of the human community and to life on earth as we know it." The mere prospect of emissions limits, he said, "shows how far the community of nations has come in accepting responsibility for

25  Ibid., 204.

26  Ibid., 205.

27  Intergovernmental Panel on Climate Change, *Second Assessment Report*, Working Group I, 1995, SPM-4.

its shared stewardship for the future of our planet."[28] President Bill Clinton said the protocol "strongly reflects the commitment of the United States to use the tools of the free market to tackle this difficult problem."[29]

Unlike the original UNFCCC, the Kyoto Protocol set binding limits. They were largely ignored. Still Republicans scuttled ratification and set a pattern generally unacknowledged by Americans: in each of the big climate agreements (Kyoto, Copenhagen, and Paris), American delegates, lobbyists, and policymakers sabotage progress at the last hour. Before blocking ratification though, Dick Cheney and his minions successfully rammed in language exempting most military activity from emissions accounting.

Alongside conservative think tanks like the Marshall Institute, the Cato Institute, and the Heartland Institute, oil industry front groups including the Global Climate Coalition (GCC) and the American Petroleum Institute organized responses of their own. Established in 1989 in the aftermath of Hansen's testimony and the IPCC, the consortium's participants over twelve years included Exxon, Mobil, Chevron, Texaco, BP, Enron, Shell, Phillips, Dow Chemical, General Motors, Ford, American Petroleum Institute, the US Chamber of Commerce, and more than two dozen others. After the *Third Assessment Report* in 2001, which included Michael Mann's now-famous "hockey stick" graph that showed skyrocketing temperatures, the GCC said international cooperation should be ditched for innovative "government/industry partnerships."[30] Public-private partnership in the guise of green goals remain a mainstay for bourgeois delay tactics. Let's rewind a few years to see how key participants were already engaged.

In their landmark study "Assessing ExxonMobil's Climate Change Communications (1977–2014)," Geoffrey Supran and Naomi Oreskes analyzed the gap between internal findings and public posturing.[31] Key perceptions affect political pressure such as

---

28 "Reduction of Greenhouse Gas Emissions Will Produce Environmental, Economic Benefits, Secretary-General Tells Kyoto Conference," United Nations, December 5, 1997, un.org.

29 "Clinton Hails Global Warming Pact," *CNN*, December 11, 1997. cnn.com.

30 "GCC Reaction to IPCC Report," Global Climate Coalition, March 2, 2001. web.archive.org.

31 Supran and Oreskes, "Assessing ExxonMobil's Climate Change Communications."

the reality of anthropogenic warming, its cause, severity, and solvability. Undercut any one of these four perceptions, and pressure to strand assets derails.

Advertorials are paid advertisements written to look like newspaper editorial pieces, giving a veneer of authority that mirrors the style of the *New York Times* and others. Advertorials the report analyzed were from Mobil before the 1999 merger and from ExxonMobil after. ExxonMobil attacked Supran and Oreskes as if this difference were unacknowledged (it was) or legally relevant (it's not). The industry would like us to believe that a small department at Exxon ran some research but this knowledge remained siloed. Not so. As early as 1983, Mobil's internal reports showed that the company understood the implications: "Some people, perhaps realistically, believe society cannot react in time to prevent major climate changes."[32] And after the IPCC published its second report in 1995, Mobil scientist L. S. Bernstein wrote a primer for the whole GCC saying the greenhouse effect was "well established and cannot be denied."

In correspondence with me, Oreskes drew a more salient point about knowledge saturation within the industry at the time:

> By the late 1980s/early 1990s, a CEO would have had to have been living in a cave not to be aware of the evolving science. Indeed, when Jim Hansen testified in Congress in 1988 that anthropogenic climate change was underway, this was reported on the front page of the NYT. It's not remotely plausible that Mobil executives were unaware of this, and if they somehow were unaware, then they would have been highly negligent, insofar as this potentially affected their core business. Various lawsuits, e.g. involving asbestos, have demonstrated that claiming ignorance is no excuse in such cases.[33]

So in the 1990s, Mobil paid the *New York Times* to run advertorials. After the 1999 merger, ExxonMobil continued the strategy. The goal of the advertorials was to sow doubt in the science or raise objections over economic consequences of mitigation.

An early advertorial in 1995 said, "The sky is not falling," while another in 1997 titled "Reset the Alarm" advised:

---

32 "Atmospheric Greenhouse Effect: Is Burning of Fossil Fuels Affecting World Climate?," Mobil Oil Corp., June 1, 1983, 3.

33 Naomi Oreskes, email to author, October 27, 2021.

Let's face it: The science of climate change is too uncertain to mandate a plan of action that could plunge economies into turmoil . . . Scientists cannot predict with certainty if temperatures will increase, by how much and where changes will occur. We still don't know what role man-made greenhouse gases might play in warming the planet . . . Let's not rush to a decision at Kyoto. Climate change is complex; the science is not conclusive; the economics could be devastating.[34]

Advertorials in 1997, the year of the Kyoto Protocol, followed the kettle logic from Freud's old joke: a man borrowed a kettle from his neighbor only to return it with a hole in the bottom. The borrower defended himself: first, the kettle already had a hole in it; second, he gave it back undamaged; third, he never borrowed the kettle!

Likewise, heating isn't happening ("Just as changeable as your local weather forecast, views on the climate change debate range from seeing the issue as serious or trivial," "high degree of uncertainty over the timing and magnitude of the potential impacts"). Or it is, but it's no big deal ("The end of the Earth as we know it is not imminent," "no longer a need for alarmists," "show a little respect for Mother Nature"). Or it's happening, but mitigation would be worse ("factory closures," "job displacement," "carbon rationing," "committing to binding targets and timetables now will alter today's lifestyles and tomorrow's living standards," "economic pain," "alternative energy approaches are not as energy efficient").

## Free Trade and the Neoliberal Turn

Perhaps we might have taken collective action demanded by the crisis had we not just suffered decades of union-busting or the destruction of the social safety net. Instead, all we can imagine is recycling and moral gamesmanship encouraged by a neoliberal framework that demands we justify ourselves and work out our atomized salvation. Immiseration of the working class happened in tandem with, and partly because of, the rise of what Barbara

---

34  Supran and Oreskes, "Assessing ExxonMobil's Climate Change Communications," 6.

Ehrenreich called the professional-managerial class, an educated group of middle managers and professionals who absorb ire that ought to target owners. By outsourcing what were once union jobs and curating flows of capital, free trade betrayed the working class and the environment simultaneously.

The UNFCCC was signed in the same year as the North American Free Trade Agreement, and the latter went into effect in the same year as the World Trade Organization was established. Its rules suppress illegal "discrimination," such as when states wish to subsidize local businesses or incentivize local labor. Some trade agreements prohibit local, lower-carbon distribution channels. Energy charter treaties allow corporations to sue governments for hurting oil revenues. The stalled Trans-Pacific Partnership originally included language protecting governments' ability to comply with the UNFCCC, but US negotiators nixed those protections. In short, it's often illegal to reduce emissions.

Worse yet, emissions accounting hides legal liability. In *This Changes Everything*, Naomi Klein explains, "Emissions from the transportation of goods across borders . . . are not formally attributed to any nation-state and therefore no country is responsible for reducing their polluting impact."[35] Under UNFCCC rules, a product consumed in Europe or the United States but produced in China or Bangladesh counts against the latter two. Globally, 22 percent of emissions stem from goods produced in one country but consumed in another.[36] In effect, emissions accounting schemes incentivize a material denial.

China is now the largest greenhouse gas emitter in absolute terms, but it is also the largest emissions exporter. If the United States had to count emissions from overseas products consumed, its emissions balance sheet would rise 6 percent. The UK's would rise a staggering 36 percent. China's emissions would fall by 13 percent, India's by 9 percent.[37]

Free trade, globalization, and neoliberalism emerged alongside climate negotiations as contradictions that needn't have been. Klein lamented, "The NAFTA signing was indeed historic, tragically

35  Naomi Klein, *This Changes Everything* (New York: Simon & Schuster, 2015), 79.

36  Zeke Hausfather, "Mapped: The World's Largest $CO_2$ Importers and Exporters," Carbon Brief, May 7, 2017, carbonbrief.org.

37  Ibid.

so . . . A new trade architecture could have been built that did not actively sabotage the fragile global climate change consensus."[38] Our pledges discourage emissions while trade agreements encourage more. Compliance in one frame compromises another.

## No Alternative to Carbon Taxes?

On a highway through southwestern Pennsylvania, I drove past a billboard with a smiling mother and child sitting in a green field. The text read "Carbon Dioxide Is Essential for Life." The advertisement was paid for by the $CO_2$ Coalition, a successor to the Marshall Institute, with funds from the Kochs and the Mercers, that communicates "the vital role carbon dioxide plays in our environment."

If "there is no alternative" to neoliberal capitalism, then there is no end to the bilkers and front groups who nudge us closer to ecocide so a wealthy few might accumulate. Universities now supplant the think tank. "You can forget about asking for money from Exxon," said one former chair of the Marshall Institute. "They send all their money to Stanford or Princeton for greenwashing."[39] Tactics evolve but the strategy cannot, because there's no acceptable alternative. Reactionaries know it. At a Heartland Institute conference, Republican representative Tom McClintock correctly identified the stakes: "If the earth truly hangs in the balance, well, then no measure is too extreme. No cost is too great. No government excess is too oppressive to enact their agenda . . . How much of a sacrifice is it if the alternative is a dead planet?"[40] To be clear, the member of Congress took the dead planet side.

If there's no alternative, the menu of market-friendly solutions constricts to carbon taxes or cap-and-trade. With the former, a tax is assessed per tonne of carbon dioxide emissions. With the latter, a cap is placed on emissions (the cap theoretically lowers over time) and companies above this cap can trade with other companies. We saw versions of these moderately effective if limited mechanisms previously with the EU Emissions Trading System and US ZEV systems.

---

38  Klein, *This Changes Everything*, 85.

39  As quoted in Kate Aronoff, *Overheated* (New York: Bold Type Books, 2021), 33.

40  As quoted in ibid., 24.

The closest the United States came to a nationwide cap-and-trade system was the Waxman-Markey bill, the "American Clean Energy and Security Act of 2009," which was ostensibly supported by an association of fossil fuel companies called US Climate Action Partnership (USCAP). Fully 9 percent of all lobbying dollars was spent on climate in 2009 and 2010. The bill passed the House of Representatives, but not before banks lobbied successfully for a carbon derivatives market. Cap-and-trade, or "crap-and-tax" as reactionaries call it, faced off against a racially charged, conspiratorial Tea Party core of the GOP. Simmering racism exploded at the election of President Obama. No poll, so far as I am aware, ever showed even half of Republican voters believed Obama was a US citizen (two-thirds to three-quarters expressed belief in the racist "birther" conspiracy theory). "The Kochs found white supremacy to be easy kindling," argued Kate Aronoff, and they pumped funding into this movement cobbled out of astroturfed organizations and authentically reactionary racism.[41] In one early victory, Republican Trey Gowdy won a primary victory in a blowout against moderate (a relative term) Republican Bob Inglis, who now tries to rehabilitate himself as an eco-conscious conservative for whoever is foolish enough to listen. In the end, lobbyists from USCAP and friends scuttled the effort. The pattern of bourgeois denial is this: position a coalition as green, work for industry-friendly provisions in case the bill succeeds, but at the same time, move to defeat it. The Waxman-Markey bill was never brought up for debate in the Senate. Obama turned his attention to health-care reform.

On the other hand, proposals for carbon taxes are frequently resuscitated. Proposals for how much to charge for a tonne of carbon dioxide range widely, from a few dollars to several thousand, with $100–$150 treated as a reasonable range that almost nobody actually pays. A lesson in bourgeois denial: fossil fuel companies call for carbon taxes.

ExxonMobil's loud support for carbon taxes crashed in spectacular embarrassment. Posing as headhunters for a major client, UK Greenpeace interviewed senior ExxonMobil lobbyist Keith McCoy on his tactics in eight years of working Congress over. "Did we aggressively fight against some of the science? Yes," McCoy confirmed. "Did we join some of these shadow groups to work against some of the early efforts? Yes, that's true. But there's nothing

41   Ibid., 87.

illegal about that. We were looking out for our investments, our shareholders."

McCoy likened chatting up members of Congress to fishing. He named Senator Joe Manchin, the West Virginian who makes a half million dollars per year from his coal company, as a key ally in weekly contact. The senator is cast as a roadblock, a conservative who is a Democrat in name only, but he angles for the same class interests represented by the party. Manchin helped scuttle both the Waxman-Markey climate bill (he ran an ad in which he shot a rifle at the bill) and Biden's Build Back Better Act, two of the most potentially transformative climate bills ever introduced in the United States. Shortly after the manuscript for this book went into production, a muted version of Biden's legislation passed, the Inflation Reduction Act, which was both a monumental achievement and a lesson on how victories are inflated; the legislation was touted by many in the media and congress as cutting emissions by 40 percent even though modeling projected a smaller cut of 5–10 percent. Finally, McCoy illuminated ExxonMobil's call for carbon taxes. "You know, nobody is going to propose a tax on *all* Americans, and the cynical side of me says 'Yeah, we kind of know that,'" McCoy mused. "No, a carbon tax isn't going to happen. And the bottom line is it's going to take political courage, political will in order to get something done, and that doesn't exist in politics—it just doesn't!"

Taxes on emissions are a political liability, as France's yellow vests demonstrated: "They talk about the end of the world and we are talking about the end of the month." As a workaround, a dividend paid out to the citizens from taxes collected could offset costs. When assessing a carbon price (set by policy or carbon markets) to mitigate emissions in its integrated assessment models, the IPCC found limiting warming to 1.5°C required a carbon price of \$135 to \$6,050 per tonne of carbon dioxide by 2030, rising to as much as \$30,100 per tonne by 2100.[42] Nowhere in the world are such high prices found.

Unfortunately, recent evidence suggests the tax and dividend scheme remains a liability even if people do benefit financially. Researchers examined two nations that implemented such a scheme, Canada since 2019 and Switzerland since 2008.[43] The

42 Intergovernmental Panel on Climate Change, *Global Warming of 1.5°C*, 2018, 152.

43 David Roberts, "Do Dividends Make Carbon Taxes More Popular?

results weren't great. In Canada, the carbon tax ($30 per tonne) was popular with liberals but not conservatives. One in six respondents was unaware they received the rebate, and conservatives underestimated the size of their rebate. In Switzerland ($105 per tonne), where the scheme is more than a decade old, 85 percent didn't know they got a refund. Perhaps this helps explain why the Swiss voted against a referendum that would have increased the tax and their rebate in 2021. Perhaps the people need to be told how greatly they benefit? No, when researchers supplied respondents with information about their benefit, respondents refused to believe it: "Canadians who learned the true value of their rebates were significantly more likely to perceive themselves as net losers even though most Canadians are net beneficiaries."[44]

Liberals fall for stalling tactics constantly as if their opponents negotiate in good faith. Without a US left party, there is no viable path for climate action that does not run through full Democratic control, yet it's unclear whether there's a path even with full Democratic control. Liberal capitalism is resilient and adapts too late to prevent misery but just in time to preserve marginal legitimacy. So we're left with small imaginations tinkering at the edges, staging a theater so all remains basically the same. Like Charlie Brown trusting Lucy will not pull the ball away this time, like Coyote chasing the Roadrunner, like Sisyphus himself reaching the top of the hill with his stone, it might be the case that neoliberal policy, public-private partnerships, and market-based solutions work out *this time*. Maybe. But I'd say the odds favor markets eyeing fuels worth a few hundred trillion dollars.

The Paris Agreement replaced mandatory emissions cuts with voluntary contributions. Here are words that don't appear in the Paris Agreement: fossil fuel, oil, coal, natural gas, renewables, green energy, solar, wind, geothermal, carbon capture, enforcement, stranding assets, capital, capitalism, carbon dioxide, methane. Greenhouse gas and emissions are terms that appear frequently, but the specifics are left vague. Words that do appear in the Paris Agreement: finance or financial (20), voluntary (6), public and private

---

Apparently Not," *Volts*, January 24, 2022. volts.wtf; and Matto Mildenberger et al., "Limited Impacts of Carbon Tax Rebate Programmes on Public Support for Carbon Pricing," *Nature Climate Change* 12 (January 24, 2022): 141–7.

44  Mildenberger et al., "Limited Impacts of Carbon Tax Rebate Programmes."

sector participation (1), economy or economies (3), economic growth (1), economic diversification (2), economic integration (12), mitigation (23), adaptation (47), migrants (1).

Days after reaching an agreement, Obama returned and lifted the oil export ban on December 18, allowing our glut of fracked shale oil to flow into the world. America is now the world's top exporter of natural gas. Obama took credit for the boom. "Suddenly America's like the biggest oil producer and the biggest gas—that was me, people."

# PART III

*Decarbonization*

# 6

# Fossil Capital

*In every stock-jobbing swindle everyone knows that some time or other the crash must come, but everyone hopes that it may fall on the head of his neighbour, after he himself has caught the shower of gold and placed it in secure hands.* Après moi le déluge! *is the watchword of every capitalist and of every capitalist nation.*

—Karl Marx

*The Earth is not being blistered because the despoilers are stupid or irrational or making a mistake or have insufficient data . . . in a profit-maximizing world it's rational for the institutions of our status quo to do what they do.*

—China Miéville

September 2018. Buried in a five-hundred-page report by the National Highway Traffic Safety Administration is a proposal to roll back vehicle fuel-efficiency standards. The agency expects global temperatures will increase 3.5°C by century's end. What stands out, though, isn't so much an agency housed under the Trump administration acknowledging mainstream consensus. No, what stuns in an otherwise dry report is a plan to raise temperatures and sea level further.

The agency proposes new standards and explains they will, if only slightly, raise carbon dioxide concentration, temperature, and sea level.[1] Older regulations cost money and ecological consequences are forgettable. Better to relax standards and reap rewards today.

---

1 "Draft Environmental Impact Statement for the Safer Affordable Fuel-Efficient (SAFE) Vehicles Rule for Model Year 2021–2026 Passenger Cars and Light Trucks," National Highway Traffic Safety Administration, July 2018, S-15.

"The amazing thing they're saying is human activities are going to lead to this rise of carbon dioxide that is disastrous for the environment and society," explains Michael MacCracken, a senior climate scientist. "And then they're saying they're not going to do anything about it."[2] Brutes salivating over the next few fiscal quarters write off their own grandchildren.

## Carbon Budget

Do you want to know why we won't meet those ambitious Paris Agreement goals? Let me show you the math for denial futures.

A carbon budget shows how many gigatonnes of carbon dioxide we can emit before we pass a temperature threshold.[3] In this book's introduction I noted the IPCC's shortcut derived from a linear relationship between carbon dioxide and heating (transient climate response) that lets us run a back-of-the-envelope climate model: for every 1,000 $GtCO_2$ emitted, surface temperature rises 0.45°C (uncertainty range 0.27°C to 0.63°C).[4]

How high could temperatures rise if we finished off all fossil fuel reserves, as we are on track to do more or less when millennials turn centenarians? That would emit more than 3,328 $GtCO_2$ and add an additional 1.5°C on top of 1.3°C already. In total 2.8°C. But burning fossil fuels is not the only source of emissions, so expect additional warming as well.

Let's take the carbon budget as it will stand in at the beginning of 2025 as our point of reference.[5] Keep in mind total anthropogenic emissions are 41.4 $GtCO_2$ per year.

---

2   Juliet Eilperin et al., "Trump Administration Sees a 7-Degree Rise in Global Temperatures by 2100," *Washington Post*, September 28, 2018, washingtonpost.com.

3   Other greenhouse gas emissions factor in, but I will follow the IPCC's simple convention of budgeting with $CO_2$ rather than $CO_2$-eq.

4   Intergovernmental Panel on Climate Change, *Sixth Assessment Report*, Working Group I, 2021, 28.

5   $400 - (5 \times 41.4) = 193$, and $1150 - (5 \times 41.4) = 943$. In the sixth report, the IPCC set a carbon budget of 400 $GtCO_2$ from the beginning of 2020 for a two-thirds chance of limiting warming to 1.5°C, or 1150 $GtCO_2$ for 2°C with an uncertainty range of ± 220 $GtCO_2$. Some carbon budgets have switched to emphasizing a slightly higher quantity for a fifty-fifty chance of meeting a temperature target, which gives us a couple

**Carbon Budget Remaining in 2025**

- 193 $GtCO_2$ remaining to limit warming to 1.5°C
- 943 $GtCO_2$ remaining to limit warming to 2°C

Figure 6.1

At the current burn rate, the budget for 1.5°C will be exhausted by 2029, maybe sooner.[6]

## Stranding Fossil Fuels

Yet the budget lures us, gestures at flexible time left, hints at a fantasy. Chase the fantasy a bit further: to keep Earth habitable, what quantity should stay underground? Add up quantities of fossil fuels, current prices, and carbon dioxide emissions if burned. Not all crude oil and natural gas is burned, since 8 percent goes to plastics (some of which is burned as trash), tar, asphalt, and lubricants, but the vast majority is burned.[7]

A caveat on my approach: coal, oil, and gas account for only four-fifths of carbon dioxide emissions. So my graph includes lines multiplying emissions by 1.25 to simulate emissions from industry and land use while depleting fuels.[8]

Separate the terms "reserves" and "resources." Reserves are fossil fuel resources we know exist and can extract with today's technology and at today's prices. Resources are total fossil fuels known or estimated to exist. Inaccessible resources can be transubstantiated to reserves if prices, subsidies, or technologies change. We won't find another Saudi Arabia, but, for example, fracking tapped previously

---

more years. Since you will encounter this after these formulas have been written, you can estimate the remaining carbon budget roughly by subtracting ~41.4 $GtCO_2$ per year since 2020. I acknowledge different groups use slightly different budgets, and here I'm using an IPCC budget along with a Global Carbon Project emissions number.

6 Be aware that while I have deferred to the last IPCC report's carbon budget, more recent research suggests revising the carbon budget down by 100 GtCO2, further cutting the window by two and a half years.

7 Murphy, *Energy and Human Ambitions*, 124.

8 For readers who want raw numbers without a 1.25 multiplication for industry and land use, I leave emissions for reserves and resources in my table. See IPCC Working Group 3 for more sophisticated long-term scenarios.

untouchable resources. In reverse, reserves might stay in the ground if material conditions change, say, cheaper renewables or infrastructure sabotage. But that's getting ahead of ourselves.

The fossil fuel industry closely tracks inventory. Figure 6.2 is derived from current resources and reserves, price averages, and emissions per source.[9] I have separated numbers for reserves and resources to show how much money would be left in the ground to meet Paris Agreement goals.

What's the upshot? Psychotic markets fret over the need to diversify with renewables for portfolio strength, because, at current production, we will burn through all global oil reserves in about sixty years, all natural gas reserves in about fifty, and all coal reserves in a hundred and forty.[10] In two centuries, we emitted 2,600 $GtCO_2$.[11] Not even halfway through the burn.

Feel the gravity of this contradiction. Stable reserve-to-production ratios over the last half century, during which we have always had three to five decades of oil and gas left, suggest a tendency to consume all reserves while developing replacements. Capitalists want collection on as much of this value as possible and then some, as resources convert to reserves. On the other hand, the total value of resources isn't collectable in full due to a range of factors, such as extraction challenges or economic lag from climate stress. But for now the incentives are wrong. To limit to 2°C, we must leave

9    Prices used are from averages over the last five years: $90 per barrel for oil, $40 per tonne for coal, and $5 per thousand cubic feet for gas. Quantities of reserves and resources from "World Energy Outlook 2021" (International Energy Agency, December 2021) and "BP Statistical Review of World Energy 2021—70th Edition" (BP, 2021). Price of coal at $90 (averaged over 11 market 2017–21) from BP and Energy Institute, "Statistical Review of World Energy," accessed May 8, 2023, bp.com. Calculations for carbon dioxide potential from "Greenhouse Gases Equivalencies Calculator —Calculations and References," Environmental Protection Agency, accessed April 29, 2022, epa.gov.

10    New technologies, discoveries, or market dynamics could of course change the reserve-to-production ration from which we get these timelines. Numbers rounded from "BP Statistical Review of World Energy 2021." Note BP does not supply exhaustive estimates for unconventional oil, so reserve-to-production ratio indicating 53.5 years is rounded up here to 60 years.

11    Rounded from 2,390 over 1850–2019, plus additional years to 2025. See Intergovernmental Panel on Climate Change, *Sixth Assessment Report*, Working Group I, 2021, 29.

Figure 6.2. **Fossil Fuels**

**Conventional and Unconventional Oil**
- Reserves: 1.75 trillion barrels
- Resources: 6.21 trillion barrels
- Value of reserves: $123 trillion
- Value of resources: $434 trillion
- Emissions from burning all reserves: 754 GtCO$_2$
- Emissions from burning all resources: 2,669 GtCO$_2$

**Coal**
- Reserves: 1.08 trillion tonnes
- Resources: 20.80 trillion tonnes
- Value of reserves: $97 trillion
- Value of resources: $1,872 trillion
- Emissions from burning all reserves: 2,147 GtCO$_2$
- Emissions from burning all resources: 41,506 GtCO$_2$

**Natural Gas**
- Reserves: 221 trillion cubic meters
- Resources: 809 trillion cubic meters
- Value of reserves: $39.02 trillion
- Value of resources: $142.85 trillion
- Emissions from burning all reserves: 428 GtCO$_2$
- Emissions from burning all resources: 1,566 GtCO$_2$

**Total Fossil Fuels**
Reserves value: $259 trillion
Resources value: $2,449 trillion
Reserves emissions potential: 3,328 GtCO$_2$
- Reserves plus one-quarter to simulate additions from industry and land use: 4,160 GtCO$_2$
- Resources emissions potential: 45,740 GtCO$_2$
- Resources plus one-quarter to simulate additions from industry and land use: 57,175 GtCO$_2$

**Reserves to be Stranded in Compliance with IPCC Carbon Budget as of 2025**
- To meet 2°C limit budget (943 out of 3,328 GtCO$_2$), we need to leave 72 percent of reserves in the ground, stranding assets worth $187 trillion
- To meet 1.5°C limit budget (193 of 3,328 GtCO$_2$), we need to leave 94 percent of reserves in the ground, stranding assets worth $244 trillion

**Resources to be Stranded in Compliance with IPCC Carbon Budget as of 2025**
- To meet 2°C limit budget (943 out of 45,740 GtCO$_2$), we need to leave 98 percent of all resources in the ground, stranding assets worth $2,400 trillion
- To meet 1.5°C limit budget (193 out of 45,740 GtCO$_2$), we need to leave 99.6 percent of all resources in the ground, stranding assets worth $2,439 trillion

72 percent of reserves in the ground, stranding assets worth \$187 trillion. For the more ambitious goal of 1.5°C, we must leave 94 percent of accessible fossil fuels in the ground, stranding assets worth \$244 trillion. Total resources in the ground, almost all of which must be stranded: \$2.4 quadrillion.

Cheap talk of vaguely "fighting" climate change assumes states and companies, all of whom already claim these assets in their portfolios, will voluntarily walk away from mountains of cash. Even as cheaper renewables come online, everyone will want a piece of the remaining mountain. A trillion here, another trillion there.

## Market Logics: Overshoot and Energy Charters

When the global economy ground to a halt to slow the pandemic, emissions in 2020 declined 5 percent, or about 2.5 $GtCO_2$ for year.[12] If we emit a little over 41.4 $GtCO_2$ per year, you can do the math. The shutdown postponed overshoot by three weeks.

Suppose those 2050 goals are flexible. Suppose self-styled pragmatists take office and tell populations: yes, the threat is real, but working families "feel pain at the pump." After all, by this point, the public will be coaxed into expecting "overshoot" beyond Paris Agreement goals but told carbon capture will bring us back down. Exploit Earth a bit longer, maybe even burn those hard-to-reach resources worth \$2.4 quadrillion.

Burning all resources would inject 45,740 $GtCO_2$ into the atmosphere, comparable to carbon dioxide levels at the end-Permian extinction. That's enough to produce a forcing of 16 $W/m^2$ and raise temperatures an additional 20°C above Earth's current cool 15°C average. Earth's surface comes up to roughly human body temperature.[13] Only very high latitudes would be habitable at all for a lucky few growing food in hell.

---

12 "Emission Reductions from Pandemic Had Unexpected Effects on Atmosphere," NASA Jet Propulsion Laboratory, accessed March 24, 2022, jpl.nasa.gov.

13 Equations are my own, derived from IPCC. Converting 45,740 $GtCO_2$ to carbon, halving for atmospheric uptake (a conservative estimate since diminishing sinks would actually leave more in the atmosphere), and using 2.13 GtC per 1 ppm, concentration is ~5,850 ppm. For forcing: $5.35 \times \ln(5,850/280) = 16$ $W/m^2$. For temperature: $45,740 \times 0.45 = 20.5$°C, though uncertainties at this absurd emissions scenario are admittedly vast. I have

This absurd scenario will not come to pass, but the logic of capital Marx identified as M-C-M′, of money transformed into commodities then back to more money through surplus accumulation from labor and exploitation of nature, will point in this direction for as long as it can. It's just a bit of overshoot, we'll hear, nothing to worry about.

Divestment signals remain mixed. Between 2012–20, fossil fuel companies lost over a hundred billion in equity and severely underperformed a key index by 52 percent. Even so, shares totaled a tenth of all equity raised in that period.[14] Holdings are concentrated in a few large companies. The world's second-largest asset manager, Vanguard, holds $300 billion in fossil fuels. The largest manager, BlackRock, owns $100 billion. Let us watch carefully for the moment they direct policymakers to dial up subsidies or nix emissions targets.

As we saw before, certain trade agreements block phaseouts. Since 1991, the Energy Charter Treaty is one egregious case. Binding more than fifty countries, the treaty enables companies to sue states that are winding down extraction.[15] Conceived to secure investments in former Soviet states, ECT arbitration can even target small "buy local" initiatives. The United States is an observer, though not a member, of ETC, but many EU states are members. Fears of lawsuits force states to reconsider policy such as when Canada-based Vermilion threatened an ECT complaint over a French plan to phase out fossil fuels. Trillions in assets will be protected by these arbitration laws.

The problem isn't individual sociopaths. What would happen if fossil fuel CEOs collectively agreed to wind down production? Overnight we would see countless startups applying for leases to extract, and governments dependent on those revenues would happily distribute them. Capitalists are products of an M-C-M′ logic. Can we decouple GDP and energy from fossil fuels in time to meet Paris Agreement goals if they provide four-fifths of all energy?

---

not included cooling effects from aerosols in this example. See also Hansen et al., "Climate Sensitivity, Sea Level and Atmospheric Carbon Dioxide."

14  "A Tale of Two Share Issues: How Fossil Fuel Equity Offerings Are Losing Investors Billions," Carbon Tracker, March 21, 2021, carbontracker.org.

15  See Fabian Flues, Cecilia Olivet, and Pia Eberhardt, "This Obscure Energy Treaty Is the Greatest Threat to the Planet You've Never Heard Of," in *Beyond the Ruins* (London: Verso, 2021).

Evidence is lacking. Can we accomplish this task within pseudo-democracies masking oligarchies, where craven policymakers fear withdrawn support from those who remember more lucrative times? Will the lords voluntarily walk away from hundreds of trillions of dollars without force? Who will force them?

## Market Logics: Greenwashing Finance

On a 1983 surf trip in Fiji, then college student Jay Westerveld snuck into a hotel to steal towels and marveled at a sign. "It basically said that the oceans and reefs are an important resource, and that reusing the towels would reduce ecological damage," he recalled. The popular resort was expanding with new construction. Surely management lost no sleep over coral. "They finished by saying something like, 'Help us to help our environment.'"

A few years later, Westerveld wrote term paper and recalled, "I finally wrote something like, 'It all comes out in the greenwash.'"[16] A magazine published his story on faux-eco products. Westerveld's neologism "greenwash" entered the lexicon.

The vast majority of consumers, from two-thirds to more than three-quarters depending on the poll, claim to care about a company's ecological reputation or say they'd pay more for ecologically friendly products. People are fickle. Who really knows what they desire?

The carton of milk I buy for my child advertises itself as "carbon positive," whatever that means. Petroleum and coal should be more rightly called carbon positive, should they not? Presumably, an ad agency guessed it sounded cheerier than "carbon negative." Even my child's milk must join the fight. What a wonderfully perverse ad.

Wall Street mirrors this scam for big investors with environmental, social, and governance funds, or ESGs. We previously saw Delta's ESG report say the airline was carbon neutral. Conservatives currently attack ESGs as "woke capital" or liberal posturing, and while their criticisms are always in bad faith, their ire contains a kernel of truth insofar as these funds are not what they claim. The problem is rather that these funds are precisely what capital needs to theatrically perform concern. To greenwash petroleum in their portfolios, firms eye ESGs and excuse bad companies like Shell

---

16 Bruce Watson, "The Troubling Evolution of Corporate Greenwashing," *The Guardian*, August 20, 2016, theguardian.com.

and ExxonMobil with supposedly good companies like Amazon. BlackRock and Vanguard have leaned heavily into ESGs. Along with State Street, these "Big Three," which own a fifth of the average S&P 500 company, shape market trends.

As Adrienne Buller explains in *Beyond the Ruins*, "ESG investors and financial products buy shares in companies (or their bonds) based on metrics purporting to measure their carbon emissions intensity, equitable labour practices, transparency, the diversity of their executive boards and so on . . . the reality is somewhat different."[17] The industry sets no parameters defining sustainability. In fact, Buller noticed, Vanguard's flagship ESG fund invests in Apple, Amazon, Microsoft, Facebook, Google, and Tesla. Buller found 809 UK funds labeled "ethical" and 150 specifically market as ESGs. Of those specifically marketed as climate-themed, a third held oil and gas companies in their portfolios.[18]

How big are ESGs as a percentage of the $140.5 trillion in global assets under management? If they grow 15 percent per year, half their recent rate, ESGs will exceed $50 trillion by 2025.[19] Before long, every sector of the market will clothe itself in green finance.

A few questions on investments as a denial instrument. First, do individuals want good returns even if it means their asset managers like BlackRock or Vanguard invest in fossil fuels? Second, when stocks in fossil fuel companies rise after COP meetings where the nations promise to zero emissions, doesn't this signal the markets interpret pledges as lies? Third, if retirement funds nosedived after divestment, how would the middle classes react?

## Market Logics: Fossil Fuel Subsidies

During a Senate confirmation hearing to join the Trump cabinet, ExxonMobil CEO Rex Tillerson took questions from Democratic Senator Jeanne Shaheen. "At this time when many of our

---

17  Adrienne Buller, "Eco-Investing?," in *Beyond the Ruins* (London: Verso, 2021).

18  Adrienne Buller, "'Doing Well by Doing Good'? Examining the Rise of Environmental, Social, Governance (ESG) Investing," Common Wealth, December 2020, iii.

19  "ESG Assets Rising to $50 Trillion Will Reshape $140.5 Trillion of Global AUM by 2025, Finds Bloomberg Intelligence," *Bloomberg*, accessed April 29, 2022, bloomberg.com.

oil companies, particularly large oil companies like Exxon, are reaping very good profits, do we really need to continue these subsidies?"

"I'm not aware of anything the fossil fuel industry gets that I would characterize as a subsidy," Tillerson responded. "Rather, it's simply the application of the tax code." He rambled about putting America first and never placing US companies at a disadvantage.

Shaheen pushed back and asked whether a 2009 agreement among the G20 to phase out fossil fuel subsidies meant we should revise the tax code. Tillerson deflected with contempt and mimed quotation marks with his fingers around "subsidies."

Everyone knows the fossil fuel industry gets subsidies, yet a CEO denied this before the Senate. What might the bad faith deflection suggest about how passive media consumers understand the industry's advantage?

Consumers don't pay the efficient cost of fossil fuels. Sixty percent of energy in United States electricity grids comes from fossil fuels.[20] Particulate matter from coal combustion penetrates the lungs and bloodstream of residents near power plants, who tend to be working-class people and people of color. Chronic obstructive pulmonary disease, stroke, heart disease, and lung cancer result. If a child grows up breathing air pollution near the plant providing me electricity to type this paragraph, I won't pay the hospital bill or funeral costs. Those charges are "externalities."

As the fifth largest oil company in the world (behind Saudi Aramco, Royal Dutch Shell, China National Petroleum Corporation, and BP, and just ahead of other familiar names like Total SA and Chevron), ExxonMobil gets a large slice of the $5.2 trillion in subsidies for the fossil fuel industry each year (recent evidence suggest it's as high as $7 trillion). That commonly cited sum came from International Monetary Fund papers indicating subsidies equal 6.5 percent of global GDP, growing at a quarter trillion dollars per year during the period studied.[21] However they aren't typically direct payments.

---

20    As of now, electricity production sources are 38 percent natural gas, 22 percent coal, 20 percent renewables, 19 percent nuclear, and 0.5 percent petroleum. See "Electricity in the US," Energy Information Administration, April 19, 2022, www.eia.gov.

21    David Coady et al., "Global Fossil Fuel Subsidies Remain Large: An Update Based on Country-Level Estimates," International Monetary Fund, May 2019. I am using the commonly cited numbers from IMF, but I should

Subsidies take several forms. Tillerson displayed contempt, not ignorance, when he pretended not to know ExxonMobil received subsidies. By deflecting to the tax code, he described a tax subsidy in order to deny receiving a subsidy. Other subsidies include federal land leases at below-market prices, roads and rails for supply chains, and permission to destroy environments. If corporations had to pay for ecological and human harm, they would fold.

The largest contributors to the $5.2 trillion in subsidies were China ($1.4 trillion), the United States ($649 billion), Russia ($551 billion), the European Union ($289 billion), and India ($209 billion). Coal and petroleum accounted for 85 percent of subsidies globally. The greatest post-tax subsidy was underpricing for air pollution (48 percent), then undercharging for global warming (24 percent), environmental costs of road fuels (15 percent), and undercharging for general consumer taxes and for supply costs (each at 7 percent). Subsidies let corporations charge less than the real social and ecological costs. ExxonMobil won't pay when Manhattan drowns or when the poor get cancer.

Aside from tax loopholes, there are two key terms: post-tax subsidies and efficient prices. Post-tax subsidies mostly mean a commodity's cost doesn't include damage caused by the commodity, also called an inefficient price. An efficient price, on the other hand, includes environmental, supply, and human costs. Environmental costs are more difficult to gauge. We could tally up medical and funeral bills easily enough for a crass "value of life," but what's the price of habitat loss or extinction?

We rightly feel anger over fossil fuel subsidies. But how would people respond to non-subsidized energy? Natural gas prices are 50–80 percent of their efficient price. Gasoline and diesel prices are below 80 percent of their efficient price, and coal is typically below 50 percent of its efficient price. Coal gets the largest share of subsidies, nearly half. Petroleum takes four-tenths and natural gas the last tenth.

Air pollution kills ten million people per year.[22] "In February 2020, the world began to panic about the novel coronavirus,

acknowledge the International Energy Agency uses a much lower number for fossil fuel subsidies, around $1 trillion, due to a narrower accounting methodology. Still, IEA found fossil fuel subsidies at their highest yet in 2022.

22   Karn Vohra et al., "Global Mortality from Outdoor Fine Particle Pollution Generated by Fossil Fuel Combustion," *Environmental Research* 195 (April 1, 2021): 110754.

which killed 2,714 people that month. This made the news," ran one comparison by David Wallace-Wells. "In the same month, around 800,000 people died from the effects of air pollution. That didn't."[23] Like the biodiversity crisis, air pollution is a quiet twin of the climate crisis crowded out of headlines by apocalyptic sentiments nobody feels for steady death.

The IMF estimated fully efficient fuel prices would drop air pollution mortalities by half, drop emissions by a quarter, and reap a $3 trillion global fiscal gain. Even if renewables are cheaper, the transition means poor returns on already-existing fossil infrastructure. So we get "path dependence." Current contingencies reflect past mistakes.

Ironically, the IMF report frequently indicates "underpricing" as a market feature. Capitalism charges less than things cost. A death drive expressed through exchange.

## Market Logics: *Après moi, le déluge!*

In the Hebrew Bible, King Hezekiah of Judah fell mortally ill, and the prophet Isaiah commanded him, "Set your house in order, for you are going to die." Hezekiah prayed fiercely, protesting he'd walked in truth and conducted himself righteously. The Lord offered a bargain: fifteen more years added to his life. As a proof the sun's shadow receded ten steps on a stairway. The world re-ordered for one old king.

Envoys from Babylon arrived in Jerusalem after hearing of the king's illness. Hezekiah showed the envoys his treasures and armor. The wise prophet Isaiah confronted the fool: "Behold, the days are coming when everything that is in your house, and what your fathers have stored up to this day, will be carried to Babylon; nothing will be left . . . some of your sons who will come from you, whom you will father, will be taken away."

At this message of ruin, King Hezekiah felt relief. "The word of the Lord which you have spoken is good—there will be peace and security in my days."

In *Capital*, when Marx said the watchword of every capitalist is "*Après moi, le déluge!*," he tapped a long tradition of callous lords unconcerned about their lessers. The phrase is attributed to

---

23  David Wallace-Wells, "Ten Million a Year," *London Review of Books*, December 2, 2021, www.lrb.co.uk.

Louis XV, whose own grandson was beheaded during the French Revolution. Contemporary versions abound. Bill Clinton said, "Adam Smith's invisible hand can have a green thumb."

Nobody is spared from the logic of a mode of production built on exploitation. As Lacan put it, capitalism is the discourse or algorithm in which, instead of communicating to another person as a person, we are forced to negotiate through cloudy signifiers. We are workers relating to other workers in an environment mediated by meaningless metrics and soul-crushing tasks, alienated with daily humiliations of demerits and debts. We would like for it to go differently, Lacan said, we'd like for the master discourse to be less all-encompassing and "to say it all, not so fucking stupid."[24]

Capitalism accumulates by exploiting workers to transform nature. To workers, money facilitates the flow of commodities, at least after prices are established through circulation. In Marx's joke, money is what allows a Bible to be transformed into brandy. For Marx, value is socially necessary labor-time. Labor congeals in materials and generates value, which is stored, by contingent social agreement, in a commodity containing both a use-value and an exchange-value. However, the laborer doesn't keep the fruit of their labor. The owner sells for more than the cost of production. Labor-time creates value, owners keep the surplus, thus exploitation. But unlike how workers think of money, for the bourgeoisie, the dynamic is reversed: commodities facilitate the flow of money. If the truth were recognized, it would feel patently absurd, so instead we believe fairy tales, myths of primitive accumulation about land and wealth owed to someone because they or their fathers were industrious value-producers, whereas we must have descended from lazy vagabonds and dull laborers. Additionally, today, we must allow the rightful lords of Earth to extract and dump carbon wherever they lease.

"To insist that large-scale capitalist society created such a metabolic rift between human beings and the soil was to argue that the nature-imposed conditions of sustainability had been violated,"[25] said John Bellamy Foster. In *Marx's Ecology*, Foster underscored a "metabolic rift" subtending the relationship between capitalist

---

24 Jacques Lacan, "On Psychoanalytic Discourse," in *Lacan in Italia, 1953–1978*, trans. Jack Stone (Milan: La Salamandra, 1978).

25 John Bellamy Foster, *Marx's Ecology* (New York: Monthly Review Press, 2000), 163.

society and Earth. Humans are materially estranged from the conditions in which we evolved. Large-scale agriculture and manufacturing feed on fossil fuels to exploit the soil. Modern agriculture needs ten times the energy inputs it gives back in food calories, a ratio made possible by fossil fuels. The rifts are everywhere.

Everywhere, yes, but one boutique industry of bunkers for the super-rich paints a comically clear picture of the sentiment expressed by *"Après moi, le déluge!"* We could call it a "bunker logic" expressing what Frank Wilhoit said of the reactionary mind: "Conservatism consists of exactly one proposition, to wit: There must be in-groups whom the law protects but does not bind, alongside out-groups whom the law binds but does not protect."

Rich reactionaries believe they can insulate themselves from the myriad horrors on the horizon. The early days of Covid-19 demonstrated their illusions as private jets scrambled toward cloisters in the Hamptons or rural Idaho. They believe they will, like Hezekiah or Louis XV, find peace in their time. At least in their compound.

Many bunkers are Cold War–era retrofitted bases or missile silos. Others are newly burrowed out. Whether obscenely luxurious or pragmatically drab, they share a fantasy of invulnerability.

A contractor called Rising S said inquiries ballooned in the first year of the pandemic. "Our buyers are looking for protection," the company said. "We are on the brink of social-civil collapse, and that is the real threat we all face." One client described his three-bedroom, two-bath, plus gun room unit stocked with Blu-Ray movies and Xbox for entertainment and tablets so kids can continue their education in case of nuclear war or a virus. "It's a whole different story when people are fighting over food and toilet paper," he explained, couching his anal concerns for toilet paper in a scenario where his family rides out societal collapse on beans and rice while playing video games and logging on for virtual school (teachers and peers left behind aren't his business). A narcissistic fantasy: in the bunker, food won't run out, solar power won't glitch, the septic systems will function, the internet will mediate needs, and the bulletproof walls will hold.

Another company, called Vivios, says its bunkers aren't your grandfather's bunker, as if that's a familiar comparison. Instead of spartan metal walls, Vivios uses ultra-modern designs with indoor pools and a simulated daytime sky. Its compound in South Dakota has a cinema and spa. The company says this largest survival community in the world is a "Safe Zone," while its comically masculine and militaristic advertising plays hilariously on cowed rich men's insecurities. How can they discipline guards if money ceases to circulate? When one broker asked, "How do I maintain authority over my security force after the event?" suggestions included locking up food or a disciplinary collar. Vivios resolves the issue by replacing guards with robotic guns.

It's all fantasy shredding surplus. Marx profoundly observed, "It is not the consciousness of men that determines their social being, but, on the contrary, their social being that determines their consciousness." At every level, denial is built into our material relations. Adorno said it best: workers have much more to lose than their chains now, namely their cars and feeble signifiers of independence or status. Right now, 80 percent of our energy comes from fossil fuels. What we need is something like war communism, a planned economy or at the very least targeted degrowth in key sectors, but what we get is cryptocurrency pyramid schemes or underground lairs suited to movie villains.

What we need, then, is decoupling in at least three ways, not necessarily in this order. We need to decouple energy from fossil fuels. We need to decouple economic growth from fossil fuels. We need to decouple GDP from resources and energy consumption. Whether this last is possible is unclear. Evidence is lacking, so we'll likely need to wind down growth due to inflexible resource, pollution, and eventually thermodynamic limits.

We will transition off fossil fuels. That much is certain. We're on track to burn through all oil and natural gas reserves around the time millennials become centenarians. Coal will hold out a bit longer. Unfortunately, since those timelines have been roughly the same for fifty years, going back to the 1980s, there is a discernible tendency of reserve-to-production ratios to stabilize, pushing the time remaining on reserves further into the future. It is hard to say how the rates of depletion will change, extend, or truncate timelines, but come the horizon will.

# 7

# Energy Trajectories

*The lot of workers today is actually no longer as it was in the classical analyses of Marx and Engels . . . the proletarians today genuinely have more to lose than their chains, namely their small car or motorcycle as well, generally speaking—leaving aside the question of whether these cars and motorcycles are perhaps a sublimated form of chains.*

—Theodor Adorno

Secretary of State Hillary Clinton launches an eight-day trip spanning Scandinavia to Turkey to coordinate with the Arctic Council. In Norway, she boards a vessel to tour northern waters.

Experts show her the ice melt. "Many of the predictions about warming in the Arctic are being surpassed by the actual data," Clinton later recalls of the conversations aboard. "That was a—not necessarily a surprising but sobering fact to be told." She calls for international cooperation on the overwhelming challenge, about which she feels optimistic. "There are things every one of us can do, and we should get about the business of doing it."

Clinton is in Norway to negotiate access to the Arctic with ExxonMobil and Norway's Statoil. She sees with her own eyes the damage and laments the "sobering" facts. But she is there to dial up the carbon dioxide. It must be done.

Three years later, in July 2015, the future Trump administration secretary of state Mike Pompeo takes the stage in a Kansas church and warns against sexual perversion and the false god of multiculturalism. He decries the Supreme Court's recent decision to allow marriage equality. "And this won't be the end of their demands, I can assure you . . . We will continue to fight these battles," Pompeo said. "It is a never-ending struggle until that moment, folks . . . until the Rapture."

Pompeo later tells a reporter the Bible "informs everything I do." Two-thirds of evangelicals in America, or around a quarter of Americans overall, believe the world will end in the next few decades.[1] A secretary of state who roots policy in the End Times is simply an average American.

Later, in his role with the Trump administration, on the day the UN announces one million out of the world's eight million species are at risk of extinction, Pompeo travels to the Arctic. "The Arctic is at the forefront of opportunity and abundance," Pompeo says from Finland. "America is the world's leader in caring for the environment."

When asked why he didn't mention climate change, he's ready. "The climate's been changing a long time. There's always changes that take place," Pompeo insists. "Societies reorganize, we move to different places, we develop technology and innovation," he says. "We will figure out responses to this that address these issues in important and fundamental ways."

Clinton's reasons were not Pompeo's, but the results were identical.

## Net-Zero Commitments and Carbon-Free Electricity

Out of the 195 signatories to the Paris Agreement, little more than a dozen enshrined net-zero requirements into law. A few small states with lots of forest cover self-declared victory. Well over a hundred pledge net-zero emissions around midcentury. The United States and the UK aimed for net zero by 2050, China for 2060, India for 2070. Turkey targeted the oddly specific 2053.[2] We'll see, in retrospect, whether pledges advance goals or stall them.

To reach net-zero emissions, we have three broad strategies: improve energy efficiency, electrify everything with carbon-free electricity, and capture remaining emissions in hard-to-decarbonize sectors. The bulk of the change will be the second strategy. Let's examine one case study that, though limping along as of this writing, might be decapitated by the time of publication.

---

1 I have previously cataloged all extant survey data on Christian apocalypticism in America. See chapter 1 of DeLay, *Against*.

2 See "Net Zero Scorecard," Energy & Climate Intelligence Unit, accessed April 29, 2022, eciu.net.

At a summit soon after his inauguration, President Biden committed the United States to net-zero emissions economy-wide by 2050. He focused heavily on jobs and workers, touting mixed metaphors ("Fuel an economy that creates jobs") and protectionism ("Make it in America"). Most importantly in the medium term was the electricity goal: "The United States has set a goal to reach 100 percent carbon pollution–free electricity by 2035."

If nuclear capacity cannot scale in so little time (currently a fifth of electricity), we can take today's renewables (also a fifth) and see how fast they'd need to grow to reach the 4.5 trillion kilowatt-hours we'll need by 2035.[3] If we'd started from the moment Biden set his goal, renewables would need to scale up almost 12 percent per year. This isn't unthinkable, but it would require enormous investments. Certainly, China is scaling up renewables at an impressive pace. But the United States, derailed by a vestigial organ for suppressing minority rights called the filibuster, is unresponsive to public demand.

If this goal isn't met, another option is simply asserting victory on paper. Companies and cities increasingly claim to be 100 percent powered by carbon-free energy ("24-7 CFE" in industry speak). The journalist David Roberts explained, "When a company or city claims to be '100 percent powered by clean energy,' what it typically means is that it has tallied up its electricity consumption, purchased an equal amount of carbon-free energy (CFE), and called it even."[4] Companies purchase a renewable energy certificate (REC), where one megawatt hour of clean energy generates one REC.

RECs started as a tool for utilities to show compliance with regulations requiring a certain threshold of electricity from renewables. But alongside that statutory apparatus, a voluntary RECs market grew to satisfy demand from corporations and government entities aiming to reduce their carbon footprint. In the voluntary market, a REC can be "unbundled," meaning the certificate representing the clean energy is sold separately from the unit of electricity. These unbundled RECs can be traded on a market. Whoever eventually retires the unbundled REC gets to claim the clean electricity generation.

---

3   "Annual Energy Outlook 2022," US Energy Information Administration, March 3, 2022.

4   David Roberts, "An Introduction to Energy's Hottest New Trend," Volts, November 12, 2021, volts.wtf.

Electricity from fossil fuels and renewable sources goes to the same grid. Yet households are sometimes offered the option of paying extra for clean energy. If they read the fine print, they may notice that what their surcharge actually goes to is REC purchases.

Many challenges plague this scheme. The cheapest REC will be associated with cheapest electricity production. If there's no timestamp on the unit of electricity's production, the market will distort the REC's price. In high winds, a REC tied to wind turbines will drop in price and swing back when wind flags. More renewables could actually exacerbate market volatility and disincentivize more renewable capacity.

These problems aren't insurmountable. But what they mean right now, since most people haven't heard of an unbundled REC, is that a mayor of a city could, in an extreme example, claim to run a city on carbon-free electricity even if none of its generation came from renewables. That's the marketing magic of an unbundled REC.

Electricity makes up 18 percent of US energy consumption. If Biden's goal were reached, fossil fuels could still dominate the energy market. More than a few consumers would feel rightly confused. To understand how these goals work, and how they play upon denial in the form of technocratic trust and optimism, we need to understand how energy works, what alternatives we have to fossil fuels, and how gaps in language are exploited as marketing gimmicks in the meantime.

## Energy and Electricity

Electricity is one slice of total energy pie. It is easy to mix up the terms. Primary energy is total energy used. End-use or consumption energy is energy used for some purpose. A lot of primary energy is lost in electricity production. For every three units of primary energy we put into electricity production, we only get one unit of electricity as end-use consumption energy (most of the loss is heat, so fewer fossil fuels would change ratios). Electricity is a little under one-fifth of global end-use consumption energy.

Energy is defined as the capacity to do work (work is the application of force through a distance). The basic unit of energy is the joule (J). Power is energy over time, such as a joule per second, and the basic unit of power is a watt (W). One joule per second is one watt.[5]

---

5 This section is heavily indebted to Murphy, *Energy and Human Ambitions*. See especially chapter 5.

The basic unit of electricity is the watt-hour, or, more commonly on utility bills, the kilowatt-hour (kWh). The term is confusing because the units look like they are energy or power, but as a convention used by electrical engineers, the kilowatt-hour is a unit of electricity. One kilowatt-hour is the amount of energy used in one hour at the power of 1,000 W, which is 3,600,000 J or 3.6 MJ.[6]

When reading about renewables and the energy transition, it is extremely common to see writers mix up terms like "energy," "electricity," and "power." With a touch of carelessness, a city's plan to generate carbon-free electricity translates into news headlines about 100 percent carbon-free energy or power. So journalists often report goals officials didn't even target. I will use these terms—"energy," "electricity," and "power," along with their units—precisely, and, once you understand them, you can see how flippantly the energy transition gets reported.

Energy comes in several forms: gravitational potential, kinetic, photon/light, chemical, thermal, electric potential, and mass (nuclear). Almost all the energy we consume through machines is essentially a heat engine with the exception of hydroelectric, wind, and solar photovoltaic. For instance, a gasoline engine converts energy in the form of heat to drive a piston, and a nuclear reactor heats water into steam to turn a turbine. Ninety-seven percent of electricity comes from a turbine (the exception is solar photovoltaic), and 84 percent of those turbines are driven by heat (exceptions are hydroelectric and wind).[7] Ninety-four percent of all energy we use is thermal.[8]

It's easy to mistake exponential growth in use of solar panels or falling prices now as implying permanently skyrocketing improvement. But there are physical limits. For instance, it will always take 4,184 J to heat one gram of water one degree. Likewise, there are limits on each renewable.

End-use consumption energy tends to be around two-thirds to at most three-quarters the size of primary energy inputs thanks to inefficiencies in electricity generation. A grid powered by photovoltaic solar would have practically no heat loss, so significantly less energy inputs could yield the same energy consumption. Figure

---

6    1 J/second × 60 seconds × 60 minutes = 3,600,000 J.
7    Murphy, *Energy and Human Ambitions*, 94.
8    Ibid., 89.

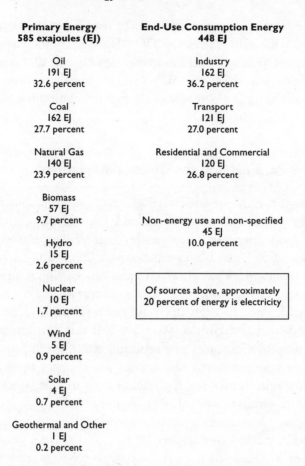

**Primary Energy
585 exajoules (EJ)**

Oil
191 EJ
32.6 percent

Coal
162 EJ
27.7 percent

Natural Gas
140 EJ
23.9 percent

Biomass
57 EJ
9.7 percent

Hydro
15 EJ
2.6 percent

Nuclear
10 EJ
1.7 percent

Wind
5 EJ
0.9 percent

Solar
4 EJ
0.7 percent

Geothermal and Other
1 EJ
0.2 percent

**End-Use Consumption Energy
448 EJ**

Industry
162 EJ
36.2 percent

Transport
121 EJ
27.0 percent

Residential and Commercial
120 EJ
26.8 percent

Non-energy use and non-specified
45 EJ
10.0 percent

> Of sources above, approximately
> 20 percent of energy is electricity

Figure 7.1. Global Primary Energy and End-Use Consumption Energy

7.1 shows global primary energy and end-use consumption energy today.[9]

Fossil fuels still dominate inputs for electricity. Numbers change quickly, but right now in the United States, electricity is 40 percent from natural gas, 19 percent from coal, and 1 percent from petroleum. The rest are alternatives, with one-fifth from nuclear and one-fifth from renewables. Of the renewables-powered electricity, wind is 8.4 percent, hydroelectric is 7.3 percent, solar is 2.3 percent, biomass is 1.4 percent, and geothermal is 0.4 percent.[10]

---

9   "US Energy Facts Explained," US Energy Information Administration, May 14, 2021, eia.gov; and Intergovernmental Panel on Climate Change, *Sixth Assessment Report*, Working Group III, 2022, 92.

10   "Electricity in the US," Energy Information Administration.

To put it all together, the world uses 19 terawatts (TW). The United States alone uses 3.3 TW.[11] Divided by the population, this means the average American consumes 10,000 W. A human body's metabolism is about 100 W, so we consume 100 times the energy a person naturally consumes. We live as royalty in other ages, and the bill will come due.

## Jevons Paradox and the Cheap Alternatives

In 1865, the mathematician William Stanley Jevons published a history of the coal-powered steam engine in Britain. He chastised those who "forget that economy of fuel leads to a great increase of consumption"[12] and later expanded, *"It is wholly a confusion of ideas to suppose that the economical use of fuel is equivalent to a diminished consumption. The very contrary is the truth."*[13] Increasing inefficiency, he argued, would not level off or depress demand. Instead, demand climbs. We now call this the Jevons paradox. Consumption increases as a rebound effect of efficiency.

Have you not heard renewables are cheaper than fossil fuels? "Today renewables are the cheapest source of power," said the International Renewable Energy Agency (IRENA) in but one recent example.[14] Mercifully, there's qualified truth here but no guarantee we avoid the Jevons paradox.

Over a couple of decades, renewable power generation grew nearly fourfold. In the last decade alone, solar photovoltaic costs fell 85 percent.[15] Onshore and offshore wind fell too, optimists cheered in victory, and IRENA's director concluded the world was "far beyond the tipping point of coal." We hope. Its fall is inevitable in the long term, yet coal consumption remains near record highs.

Costs of renewables are tricky to frame fairly due to massive up-front investment in competition with cheap or already existing fossil fuel infrastructure. Energy specialists use the term "levelized

---

11   Murphy, *Energy and Human Ambitions*, 202.

12   William Stanley Jevons, *The Coal Question*, 2nd ed. (London: MacMillan and Co., 1866), x.

13   Ibid., 123. Italics in original.

14   Jillian Ambrose, "Most New Wind and Solar Projects Will Be Cheaper than Coal, Report Rinds," *The Guardian*, June 23, 2021.

15   Intergovernmental Panel on Climate Chance, *Sixth Assessment Report*, Working Group III, 2022, 11.

cost of electricity" (LCOE) or a ratio of lifetime costs to lifetime electricity generation. In the Energy Information Administration's definition, LCOE "refers to the estimated revenue required to build and operate a generator over a specified cost recovery period."[16] LCOEs let us compare apples to oranges, so to speak, such as the cost of solar panels for a completely off-grid house ($50,000 in installation plus another $50,000 in batteries needing replacement every ten to fifteen years) with a tonne of coal that costs little but adds up over time. One caveat is that LCOE can be the cost only to break even, whereas utilities costs include profit, so LCOE may appear lower than a market rate. Armed with LCOE scores, media headlines become canon: "Most New Wind and Solar Projects Will Be Cheaper than Coal, Report Finds."

Electricity costs hover in the vicinity of $0.10–0.15 per kilowatt-hour, and the average US house consumes 30 kWh per day. Granted, the average American consumes energy at a pace five times the global average. Figure 7.2 on the following page shows levelized costs of electricity generation.[17]

When compared against renewables, using either the LCOE or the actual utilities price customers pay is a methodological choice. On a bill, we might pay only for ongoing operations plus a surplus for profit. Natural gas floats around $0.06 per kWh. Coal costs as little as $0.03. Fossil fuels can match or beat renewables based on marginal operating costs while losing on LCOE. We don't pay the real cost of electricity, which would include LCOE as well as social costs (externalities). The National Academies of Science estimated the real cost of energy is 170 percent higher than our bills suggest. But with LCOE scores accounting for all costs, renewables are indeed cheaper. Importantly, LCOE scores do not include costs of grid integration.[18] The paradox: yes, renewables are cheaper; yes, they may cost more on a utility bill due either because a consumer must pay for a REC or else a utility may pass integration costs along to a consumer.

---

16  "Levelized Costs of New Generation Resources in the Annual Energy Outlook 2022," Energy Information Administration, March 2022, eia.gov.

17  "Renewable Power Generation Costs in 2020," International Renewable Energy Agency, 2021, irena.org, 11; and Douglas Ray, "Lazard's Levelized Cost of Energy Analysis—Version 13.0," Lazard, 2019, lazard.com, 7.

18  Intergovernmental Panel on Climate Change *Sixth Assessment Report*, Working Group III, 2022, 12.

| Source | Cost USD/kWh (2020) |
|---|---|
| Bioenergy | $0.08 |
| Geothermal | $0.07 |
| Hydropower | $0.04 |
| Solar Photovoltaic | $0.06 |
| Solar Thermal | $0.11 |
| Onshore Wind | $0.04 |
| Offshore Wind | $0.08 |
| Coal | $0.11 |
| Natural Gas | $0.18 |
| Nuclear | $0.16 |

Figure 7.2. LCOE Comparison

Renewables are also subsidized in ways difficult to compare with fossil subsidies. Recall our $5.2 trillion per year in fossil fuel subsidies, most of which is uncompensated human and ecological damage. After that, the remaining subsidies top $600 billion. Two-thirds goes to fossil fuels, one-third to renewables, and the remainder to nuclear. Two-thirds of US energy subsidies go to the renewables making up only an eighth of our primary energy. Enormous subsidies prop up renewables. As a result, some homeowners can afford to install solar panels with tax rebates.

We should hope to see more of it. We must watch vigilantly, for these costs are not handed down by the invisible hand of the market but by policy decisions. If fossil capital feels threatened, it can tamper with subsidies.

## Alternative Renewables and Challenges: Solar

Renewable energy sources draw on Earth's energy budget, and that budget is not unlimited.[19] Of the 174,000 terawatts of radiation continually striking the top of the atmosphere, subtracting for radiation deflected by the albedo effect, 123,000 TW is absorbed by the lower atmosphere (40,000 TW) and the surface (83,000 TW). Beyond what is yielded from splitting atoms, those latter two numbers are what humans have to work with plus minor additions available from geothermal (44 TW) and tidal energy (3 TW). Until we work out fusion, no other energy sources exist.

---

19   Energy budget from Murphy, *Energy and Human Ambitions*, 175.

Solar radiation divides into the hydrological cycle (44,000 TW), wind (900 TW), photosynthesis (100 TW), and ocean currents (5 TW). Suppose we coated the entire surface of Earth with solar panels to capture all 83,000 TW from sunlight striking the surface. Capturing it all would consume energy needed for evaporation and plant life. Fortunately, we don't need so much. Civilization runs on 19 TW, which photovoltaic panels could harvest by covering half a percent of Earth's land.[20]

Photovoltaic panels draw energy from photons. Two or more adjoining slabs of silicon contain deliberately introduced impurities so one has a positive charge and the other a negative. The two panels meet at a junction where an electron from the positive side jumps to the negative, contributing to a current driven through a circuit. When photons from the sun hit the top silicon layer, they knock electrons and some jump the junction. But not every photon from the sun can push an electron over the junction. Much of the infrared passes through without absorption, and higher frequencies create heat. So fewer than half the wavelengths from the sun are effective.

Of total energy striking the panel, 32 percent can be collected, a number known as the Shockley-Queisser limit. That's a limit imposed by physics. In practice, solar panels tend to be 15–20 percent efficient.[21] All that energy goes to electricity, which isn't bad if fossil fuel–fired electricity plants convert only 38 percent of energy into useful electricity.

Another option is solar thermal (concentrated solar). Mirrors concentrate sunlight upon a container holding liquid such as synthetic oil. The heated fluid runs through a boiler to turn water into steam turning a turbine. Efficiency for solar thermal is also 15–20

---

20  In the following sections, I use 19 TW (primary energy) instead of end-use consumption energy, which is about 13 TW and is what renewables would actually need to satisfy if replacing fossil fuels today. My reasoning is somewhat arbitrary, but since we are discussing an energy system that continues to grow at 1.5 percent per year (on pace to double in forty-six years), 19 TW seems like an appropriate number both as our current primary usage and where end-use consumption will be by midcentury.

21  Expensive multi-junction PV cells exist that achieve efficiencies closer to 50 percent. This section focuses on commercially available panels due to costs, but using multi-junction PV cells suggests we might boost efficiency. Description of solar panels from Murphy, *Energy and Human Ambitions*, 212–14.

percent, and it is on the more expensive side of renewable-sourced electricity.

Photovoltaic panels have downsides. For plummeting costs we can thank outsourced manufacture to regions with low wages and few worker protections. Pointing incentives in the right direction ecologically can mean worse exploitation. Often, materials are extracted from the land under awful conditions. In his criticism of green growth and equitability, Derrick Jensen reminded us "mining is such a horrid existence that it was one of the first forms of human slavery."[22] Production can also be harsh. Four-tenths of panels are made by a few Chinese firms, some of which face serious allegations around labor conditions. Government subsidies allow firms to flood the market and sell for less than commodities cost to make, circumventing trade rules with an unfair advantage disincentivizing domestic production. Additionally, while costs have dropped 85 percent in a decade, prices are leveling out (let us hope to be surprised). Finally, while panels can be recycled, many of the components are difficult or expensive to recycle and end up in landfills.

The most commonly cited problems are intermittency of the sun and the costs associated with installation and batteries. It's not exactly a surprise. "There are lots and lots of interesting discussions to be had about the variability of renewable energy and how to accommodate it and ensure reliability," journalist David Roberts said in a barb for solar skeptics. "None of them begin with the assumption that your interlocutor doesn't know the sun goes down."[23]

Still, demand mismatches supply. Demand peaks in the evening, and cloudy winter months prevent collection. And solar panels are rated for peak capacity—that is, the wattage delivered in the four or five hours of peak sunlight. We need significantly more capacity than we use, so as to store surplus, since with solar energy, unlike with natural gas or coal-fired electricity, we cannot turn up the sun to meet demand.

Costs remain startling, especially when we hear how cheap solar photovoltaic is in LCOE terms. In a completely off-grid setup, a typical household drawing 30 kWh electricity per day would need about three days of storage in case of poor weather. Installation

22   Derrick Jensen, Lierre Keith, and Max Wilbert, *Bright Green Lies* (Rhinebeck, New York: Monkfish Book Publishing Company, 2021), 312.

23   David Roberts, Twitter, February 18, 2022, twitter.com.

of panels to draw 30 kWh would run around $50,000, and if this off-grid house used Tesla Powerwall 2 units for three times daily storage, the home would need to budget another $50,000-plus for batteries.

Very few people will live off-grid, so battery storage will be built into grids. Installation and storage costs for fifty peak terawatts, the capacity we'd need to meet demand accounting for 20 percent efficiency, adds up to $50 trillion. That's about three decades of the global petroleum budget.[24]

We must dump trillions of dollars into these projects. Let us be honest. We are indeed speaking of trillions. Reactionaries will pounce on our hesitance and say survival is too costly.

## Alternatives and Challenges: Wind, Hydro, Biomass, Geothermal, and Nuclear

Energy return on energy investment (EROI) is a ratio between energy collected and energy spent. If it takes one unit of energy spent to yield X amount of usable energy, the EROI ratio is X:1. The higher the X, the better the yield. Lower X values are tolerable if the energy is scalable (solar photovoltaic), and sometimes higher X values aren't the panaceas they appear to be, on account of severe limits to scaling (hydroelectric dams). Nuclear fission EROI is surprisingly low, due to enormous energy inputs.

EROI ratios change. As oil becomes more difficult to reach, it takes more energy to extract. At 16:1 today, oil's EROI has dropped precipitously from the early gushers that surpassed 100:1. The ratio shrinks when extraction takes more energy. Interestingly enough, in the United States, the food industry spends ten times the amount of energy contained in the final food products we consume, so our food production is heavily dependent on fossil fuels. In any prior stage of development, an EROI of less than 1:1 would have meant famine. Figure 7.3 on the following page orients us on alternatives in this section.[25]

---

24   Murphy, *Energy and Human Ambitions*, 220.

25   Mason Inman, "How to Measure the True Cost of Fossil Fuels," *Scientific American*, April 1, 2013, scientificamerican.com; Mason Inman, "Behind the Numbers on Energy Return on Investment," *Scientific American*, April 1, 2013, scientificamerican.com; and Murphy, *Energy and Human Ambitions*, 242, 244.

| Source | EROI |
|---|---|
| Hydroelectric | 40:1 |
| Wood | 30:1 |
| Wind | 20:1 |
| Coal | 18:1 |
| Oil | 16:1 |
| Sugar Cane Ethanol | 0.8–10:1 |
| Natural Gas | 7:1 |
| Solar Photovoltaic | 6:1 |
| Soy Biodiesel | 5.5:1 |
| Nuclear Fission | 5:1 |
| Tar Sands | 3–5 |
| Corn Ethanol | 1.4:1 |

Figure 7.3. Energy Return on Energy Investment

Wind turbines harvest kinetic energy from air. Blades rotate a shaft and turn an electric generator. An update to an old technology. We calculate harvestable energy with blade radius, efficiency, and wind speed. The physics says the upper limit a turbine can harvest, the Betz limit, is 59 percent, and in practice, turbines capture roughly half of the wind's energy. Only certain locations are windy enough to be worth the investment, and since wind isn't always blowing, turbines operate at about a third of capacity.

Turbines must be spaced out several diameters to maximize harvest. If we packed the maximum number of turbines possible across the United States such that they were always towering overhead, they could produce 0.7 TW of the 3.3 TW we use. Maybe a bit more if we stacked the coasts with offshore farms. In more practical estimations, the world might be able to pull 1 TW of its 19 TW.[26] Significant but limited when compared to solar.

Hydroelectric dams harvest energy from a reservoir of water. Gravitational potential energy creates pressure used to rotate turbine blades. Dams can capture an incredible 90 percent of potential energy. Drawbacks include habitat destruction and harm to animal life. There are also only so many places we can build and could theoretically harvest 2 TW out of our 19 TW, but given economic constraints, 1 TW is a more realistic upper potential. Right now, the world draws about a half a terrawatt from hydroelectric dams, meaning ambitious (and destructive) projects could double

26  Murphy, *Energy and Human Ambitions*, 193–203.

the yield. As with wind, every little bit helps, but neither wind nor hydro can cover energy needs like solar.

Biomass such as wood was the traditional source of energy for societies until the Industrial Revolution. Much of the world still relies on wood and other types of biomass (6 percent of global energy), and biofuels are increasingly important to the developed world. One form of carbon capture and storage we will examine later, bioenergy carbon capture and storage (BECCS), depends on burning biomass.

In the United States, ethanol is the primary biofuel. Since they require processing, biofuels have lower EROI than directly burning biomass. Ethanol produces the exact same amount of carbon dioxide per unit of energy as petroleum, but it's considered carbon neutral since the plant took carbon from the atmosphere. Earth's biological scale is perhaps 100 TW, and accounting for inefficiency, drawing 19 TW might require burning up all life on the planet. Prefer to avoid.[27]

Geothermal energy is another small player that depends on location. Technically, anywhere can be a source of geothermal energy, since heat flows throughout Earth's crust at 44 TW, and much of that energy could be harvested by drilling cores. Practically it's only efficient to collect geothermal energy near vents and geysers. About 13 GW of global electricity comes from geothermal energy, and the United States accounts for a quarter. In seventh place is Iceland, which gets 30 percent of its electricity from geothermal. The largest carbon direct air capture machine in the world is based in Iceland and runs on geothermal, which is why direct air capture is affordable in that specific location. Thirteen gigawatts is a minute fraction of 19 TW.[28]

Finally, we arrive at the nuclear option, which I address in fear and trembling. Nuclear fission is the alternative that isn't renewable, and like all alternatives to fossil fuels, it is not quite carbon-free given resource inputs. Nuclear energy is often a litmus test. For some, a refusal to endorse nuclear energy is tantamount to climate denial. For others, an endorsement is ecocide. For my part, I'm nuclear-friendly if not bullish, and I want to underscore I find bullish proponents obnoxious, inasmuch as nuclear capacity, which only delivers 2 percent of global energy, won't scale at the level we need by midcentury.

27  Ibid., 235, 244.
28  Ibid., 286–9.

When a neutron bumps into and sticks to a nuclide that can be used for a nuclear fission reaction (uranium-233, uranium-235, and plutonium), pieces fly away and bump into other materials and transfer energy. Uncontrolled, this reaction explodes. If, instead, the process is managed with absorbing control rods, a chain reaction is sustained. Heat makes steam. The rest of the steam turbine works much as any other heat engine.

Globally, around 450 nuclear reactors have an operating capacity maxing near 400 GW. The United States leads, followed by France, China, Russia, Japan, South Korea, and India. Plants last fifty or sixty years. The median US plant is thirty years old, and there's significant political resistance to new construction. Instead of the Cold War–era promise of nuclear-powered electricity "too cheap to meter," the Chernobyl and Three Mile Island disasters turned fission politically toxic. Plants are expensive and take large energy inputs, which is why EROI is so low. Nuclear fission costs $10 per watt compared to solar's $2.50 per watt.

We have limited reserves of uranium. The current use rate will burn through uranium reserves in ninety years. Running all 19 TW on nuclear power will use up reserves in four years. A few options remain on the table. A breeder reactor converts fissile uranium into fossil plutonium. It could extend our reserves of fissile material to centuries or as much as a millennium of 19 TW consumption. The downside is a proliferation of radioactive materials.

Nuclear reactors work by fission, or atom splitting, the same process by which the bombs dropped on Hiroshima and Nagasaki worked. Because of the United States' crime, this is the image people have in their head of nuclear power. But not long after, we built hydrogen bombs, fusing atoms rather than splitting them. A fusion reactor would free us from rare uranium, only 0.72 percent of which is fissile, since hydrogen is abundant. These types of fusion reaction would work by combining deuterium (an isotope of hydrogen) or tritium (derived from lithium, so a source of competition for lithium batteries). We would have enough deuterium to last for sixty billion years.[29]

Aside from serious ecological and social danger from radioactive materials, the problem with fusion is it's always thirty years away. For decades, we've heard fusion electricity is just over the horizon. Should we solve the technological problem, the political

29  Ibid., 249–78.

counterpressure returns. Germany recently shut down its remaining nuclear power plants, while China is reportedly opening 150 new plants in fifteen years. The United States is building two as of this writing, and the last to enter operation was in 2016. Several countries are working toward next-generation nuclear plants. James Hansen has argued nuclear and coal with carbon capture might be necessary to support base power energy needs. But nuclear power won't scale rapidly enough to cover phaseouts of fossil fuels.

## Energy Density and Batteries

Renewable energy still needs storage. Alternating current (AC) can't be stored as such but can be used to charge a direct current (DC) battery, stored mechanically (pumping water uphill then releasing downhill to charge turbines), or stored in super capacitors. Solar power and wind turbines do not store energy.

Fossil fuels are incredibly energy dense and store that energy practically forever (or at least for tens and hundreds of millions of years). The transition depends on batteries currently showing an energy density 1/46th as efficient as gasoline. Figure 7.4 on the following page gives an idea of the challenges for energy density.[30]

You find overly bearish skeptics and bullish proponents on battery progress. There are obviously physical limits to energy density in a battery, but it isn't clear where those limits are. Over four decades, the lithium battery reached as high as 1 MJ/kg, and one energy researcher put the challenge as such: "When you get the really serious battery guys over a beer and ask them, off the record, 'Tell me the truth. Has anyone you know in any of the formulations had a breakthrough?' The answer is 'No.' No one even has one on the horizon."[31] But even if breakthroughs are few, it doesn't mean there are no improvements. Lithium-ion batteries prices have fallen sharply, 85 percent in a decade, and energy density has more than tripled since the early nineties, though the trend in progress appears to be leveling off.[32]

---

30  Murphy, *Energy and Human Ambitions*, 127; Jensen, Keith, and Wilbert, *Bright Green Lies*, 193,174; and Ethan Boechler et al., "Energy Density," Energy Education, December 20, 2021, energyeducation.ca.

31  Steven LeVine as quoted in Jensen, Keith, and Wilbert, *Bright Green Lies*, 177.

32  Micah S. Ziegler and Jessika E. Trancik, "Re-Examining Rates of Lithium-Ion Battery Technology Improvement and Cost Decline," *Energy & Environmental Science* 14, no. 4 (April 21, 2021): 1635–51.

| Source | MJ/kg |
|---|---|
| Apple | 3 |
| Wood | 16 |
| Fat | 37 |
| Rocket Fuel | 17 |
| Coal | 24 |
| Jet Fuel | 43 |
| Gasoline | 46 |
| Natural Gas | 55 |
| Uranium-235 | 3,900,000 |
| Supercapacitor | 0.01–0.04 |
| Lead Acid Battery | 0.1–0.2 |
| Lithium Ion Battery | 0.04–1.0 |
| Tesla Powerwall | 0.4 |

Figure 7.4. Energy Density

What does this mean for airlines? With an upper-range lithium-ion unit storing 0.7 MJ/kg, a Boeing 737 would require twenty times as much weight in batteries to reach the same range it goes on jet fuel. This would exceed the 737's carrying capacity several times over. Or if all the space currently used for jet fuel were packed with an equal weight of lithium-ion batteries, the 737's range would drop to 200 km.

Hydrogen fuel cells are a possible alternative, though batteries still outperform hydrogen fuel cells as a storage device. Hydrogen fuel cells are a carrier or store of energy. Airplanes powered by hydrogen fuel cells would still produce greenhouse gas emissions if the cell were created with natural gas. Their efficiency remains low as well. Volkswagen found its battery-powered cars had an overall efficiency of 76 percent, while hydrogen reached only 30 percent.[33]

In fall 2021, the US startup Wright Electric Inc. became the latest to throw their hat in the ring with electric planes. The company planned to retrofit a small BA3 146 aircraft with hydrogen-powered electric motors and turn it into a zero-emissions jet. If successful, the electric plane could carry one hundred people for one hour of flight time. This is roughly in line with other companies building electric planes. Carrying small numbers of people fewer than a thousand miles is the first step. However, Wright Electric acknowledged the technology

---

33 "What's More Efficient? Hydrogen or Battery Powered?," Volkswagen, November 7, 2019, volkswagenag.com.

is not yet there and will require advances in hydrogen fuel cells or aluminum fuel cells.[34] For now, lithium batteries are what we have.

Exploited workers mine lithium from their land. Bolivia, Argentina, and Chile together have half the world's lithium. The rare earth mineral is plentiful enough that we won't soon deplete reserves, but exploitation and violence will be distributed unevenly in areas with deposits. Bolivia has the world's largest lithium deposit, at Salar de Uyuni. The area will be stripped in two decades. Combined with water scarcity artificially driven by redirection from agriculture to mines, locals will be left a ruined landscape and a dying economy.

Two points of good news. First, we aren't likely to run out of lithium soon. Second, production is currently shifting away from brine extraction in the Andes and Congo to spodumene mined under better conditions especially out of Australia. Western Australia produces half the world's lithium.[35] Skeptics say there won't be enough lithium or its humanitarian problems are endemic, and perhaps they will prove correct in the long run, much as the depletion of conventional crude oil turned to dirtier options in desperation. For now, though, our options look decent.

Reserves currently stand at twenty-two million tonnes, total resources at eighty-nine million tonnes.[36] In first place stands Bolivia with twenty-one million tonnes in lithium resources. Next comes Argentina with nineteen, Chile ten, the United States with nine, Australia with seven, and China with five. All others have less than five million in known resources. Sixty-five percent goes to batteries. If global production, currently hovering around 100,000 tonnes per year, grows 15–20 percent as it has in several recent years, we'll run out in the latter half of this century. But such growth is highly unlikely. The IEA projects lithium use will double and then level out over the next decade, but in a high sustainable growth scenario, lithium demand could grow by forty times by 2040.[37] So

---

34   Charlotte Ryan, "L.A. Startup Wright Targets 100-Seat Electric Plane by 2027," *Bloomberg*, November 4, 2021, bloomberg.com.

35   Graham Readfearn, "A Cut-and-Paste Attack on Electric Vehicle Batteries and Renewables Is Spanning the Globe," *The Guardian*, March 30, 2022, theguardian.com.

36   Davide Castelvecchi, "Electric Cars and Batteries," *Nature*, August 17, 2021, nature.com; and "Lithium," US Geological Survey, January 2022, pubs.usgs.gov.

37   "Mineral Requirements for Clean Energy Transitions," International Energy Agency, accessed May 14, 2022, iea.org; and "Committed

it is hard to say. Recycling is on the horizon, perhaps, but lithium batteries prove far more difficult to recycle than lead acid batteries.

## Climate Colonialism

In November 2019, Bolivian President Evo Morales was forced from office in what he called a lithium coup. If wars were fought over oil reserves in the twentieth century, expect conflict over lithium in the twenty-first. When asked whether the US government supported a coup so that Tesla could extract more lithium, Elon Musk boasted, "We will coup whoever we want! Deal with it."

It was a proud declaration of climate colonialism over Twitter from a man who represents a false hope of decarbonization without sacrificing lifestyles. If developed states pursue resource-rich life-styles without sacrifice, leaders will find reasons to treat people who live on top of the minerals as threats. But unlike oil-rich states, which the Global North clumsily accuses of terrorism, the villag-ers of Salar de Uyuni might one day be called enemies of the earth, threatening civilization itself. Such will be those who stand in the way of *our* lithium.

Warning of what they called the "great climate migration" over the next few decades, the philosophers Olúfẹ́mi O. Táíwò and Beba Cibralic called for climate reparations and welcome for refugees. The movement of millions internally and across borders is now certain by midcentury. The refusal to accept migrants will fire up eco-fascists, who will delight in violence, but without aid, the mass famines and resource wars will carry on in mundane violence, out of view of Western eyes. An alternative to reparations is preferred by Musk and all those who insist, at all costs, our energy-rich lifestyles are nonnegotiable. "The great climate migration that will transform the world is just beginning," warned Táíwò and Cibralic. "There are two ways forward: climate reparations or climate colonialism."[38]

---

Mine Production and Primary Demand for Lithium, 2020–2030," Interna-tional Energy Agency, accessed May 14, 2022, iea.org.

38   Olúfẹ́mi Táíwò and Beba Cibralic, "To Address the Great Climate Migration, the World Needs a Reparations Approach," *Foreign Policy*, October 10, 2020, foreignpolicy.com.

# 8

# Decoupling and the Limits to Growth

*We do not know how much more of natural land cover and what share of biodiversity we can lose with relative impunity, and if the global population were to survive in relatively large numbers for much longer than its recorded period (that is, longer than 5,000 years), it would almost certainly run into some material restrictions. And we may not succeed in this unprecedented effort to reconcile planetary constraints with human aspirations (or should it be delusions?).*

—Vaclav Smil

*The ideas of the ruling class are in every epoch the ruling ideas.*

—Karl Marx

Just after the release of the IPCC's latest report, the Biden administration moves to lease eighty million acres for oil and gas drills in the Gulf of Mexico. It's the largest fossil fuel lease in US history. As the auction moves forward in September 2021, the administration's decision draws attention:

On August 9, 2021, the Intergovernmental Panel on Climate Change released a new report detailing observations of a rapidly changing climate in every region globally. This report does not present sufficient cause to supplement the [environmental impact statement], at this time. . . . The report as well as additional analysis of climate change may be a significant consideration in the Department's decisions regarding oil and gas leasing programs in the future.[1]

---

1   Laura Daniel-Davis, "Record of Decision for Gulf of Mexico Outer

The official names the IPCC report only to ignore it while gesturing at reconsideration later.

The journalist Walker Bragman breaks the story and quotes Earthjustice attorney Brettny Hardy asking why the administration doesn't halt the auction and demand an adequate environmental study. "Why rely on five-year-old environmental analyses to continue this permitting when the law is clear [that] you've got to have updated analysis and rely on the best available science?"[2]

On the evening of the revelation, Michael Mann, one of the most well-known public climatologists in the world, responds not with criticism of Biden but by attacking David Sirota, the founder of the media outlet that broke the news. Linking to a profile characterizing Sirota as critic of mainstream media and a supporter of Bernie Sanders, Mann says, "This is hardly an objective news source."[3] A couple of days later, on social media, he again complains that "this is crap."

That night, Sirota takes notice of a climate scientist defending an administration consciously ignoring the IPCC. "You pretending this didn't happen is destructive," Sirota says of Mann. "You are prioritizing your own political access over facts."

Mann accuses the journalist of engaging in "new climate war" tactics he's cataloged in a recent book, which is critical of the left and praises what he sees as pragmatism on the right. Much of the book recounts Twitter fights like this one, but a main climate solution is voting blue, specifically electing Biden. So Mann defends the new administration. "What they said is that the report WILL impact their decision-making moving forward." He does not explain why he changed the administration's word "may" to his word "WILL."

Strange to see an important public voice, whose influential "hockey stick" graph of temperatures made the crisis so stark, defend oil rigs.

---

Continental Shelf Oil and Gas Lease Sale 257," Bureau of Ocean Energy Management, August 31, 2021, boem.gov.

2   Walker Bragman, "'Does Not Present Sufficient Cause,'" *The Lever*, September 17, 2021, levernews.com.

3   Marc Tracy, "The Former Journalist Who Is Bernie Sanders's Media Critic," *New York Times*, January 21, 2020, nytimes.com.

## Can We Decouple Production and Emissions?

In her righteous castigation of leaders at the 2019 UN Climate Action Summit in September, the line everyone remembers is "How dare you!" But Greta Thunberg hinted at a far more important conversation about "fairy tales of eternal economic growth." Depending on who you ask, it's either a truism or trite cliché to say we cannot have infinite growth in a world of finite resources. True insofar as it goes, but one can still have an awful lot of growth. Where are the limits?

If it were possible to navigate green or sustainable growth, as opposed to a steady state economy or degrowth, we'd need to decouple growth and emissions. Ever since Keeling first took measurements at the Mauna Loa Observatory, atmospheric carbon dioxide concentration and GDP have shown a tight correlation coefficient of 0.958. A growing economy means higher carbon dioxide concentration. I think of this ghastly relationship whenever I hear the economy is doing well.

For nearly a century, GDP has been the ultimate measure of "growth" in "the economy," which is more or less reflexively treated as an object that exists. In theological terms, our economy works like a theodicy, an explanation for why some go hungry while others are full. This economy, like a god, distributes resources according to a rubric of blame or deservingness. The more one's work contributes to GDP, the more righteous it must be regardless of use value to humans. A limited tool, GDP transforms all kinds of exchange into numbers seen as both mathematically and morally objective. GDP is "good" or "bad" depending on whether it grows or stalls. But why should we desire permanent growth? GDP counts nominal exchange values but not human flourishing. Volunteering my time cleaning up a park does not count toward GDP; a ghastly car crash on my way home would contribute greatly through insurance payouts and medical bills. A book study does not contribute; a hedge fund does. GDP mystifies relations of exchange and suffering. To speak of decoupling GDP from energy or emissions is already a mystification and designates little beyond underlying material relations; for example, if carbon emissions and GDP are coupled, it is because wealthier people emit more. Whether or not we still use this recent, crude metric for righteousness a century from now, it is presently a tale condoning the damnation of the Earth and its inhabitants.

Decoupling refers to breaking a relationship between economic growth and resource consumption (often expressed as emissions but also as material throughput). In practice it implies several entangled relationships between production, emissions, energy, and fossil fuels. In a meta-analysis of decoupling literature, the researchers Helmut Haberl et al. described the field as such:

> Strategies toward ambitious climate targets usually rely on the concept of 'decoupling'; that is, they aim at promoting economic growth while reducing the use of natural resources and [greenhouse gas] emissions. GDP growth coinciding with absolute reductions in emissions or resource use is denoted as 'absolute decoupling', as opposed to 'relative decoupling', where resource use or emissions increase less so than does GDP.[4]

Emissions in thirty-two mostly developed countries have peaked. US emissions peaked in 2005 and have since declined by a fifth.[5] Capitalist climate governance suggests this is proof absolute decoupling is well underway thanks to the market.

Shortly after he left office, President Obama jumped into the decoupling debate with a claim that underscores a nascent type of denial. Obama said the "United States is showing that [greenhouse gas] mitigation need not conflict with economic growth." The numbers, he claimed, backed him up. He cited a 9.5 percent decline in energy sector emissions and 10 percent economic growth over his tenure. "This 'decoupling' of energy-sector emissions and economic growth," he insisted, "should put to rest the argument that combatting climate change requires accepting lower growth or a lower standard of living."[6]

Among the thirty-two "peak-and-decline" states, about half is due to less fossil fuel use as a share of total energy. Another third

---

4  Helmut Haberl et al., "A Systematic Review of the Evidence on Decoupling of GDP, Resource Use and GHG Emissions, Part II: Synthesizing the Insights," *Environmental Research Letters* 15, no. 065003 (June 11, 2020).

5  "Inventory of U.S. Greenhouse Gas Emissions and Sinks," Environmental Protection Agency, accessed May 1, 2022, epa.gov; and Zeke Hausfather, "Analysis: Why US Carbon Emissions Have Fallen 14% Since 2005," Carbon Brief, August 15, 2017, carbonbrief.org.

6  Barack Obama, "The Irreversible Momentum of Clean Energy," *Science* 355, no. 6321 (January 9, 2017): 126–9.

is reduction in energy use due to low economic growth, so they don't exactly show what Obama hoped to prove.[7] Moreover, in peak-and-decline states, the rate of decline in greenhouse gasses is incommensurate with Paris Agreement goals. If decoupling is possible, it won't occur quickly enough.

What do we make of Obama's claim that emissions dropped as the economy boomed, the fluke of recovery from the Great Recession notwithstanding? Under the UNFCCC, emissions are counted within national boundaries (territory-based accounting), not counted against the nation that consumes the product (consumption-based accounting), so outsourced production outsources emissions. But this isn't the biggest reason emissions in the developed world declined. A full third of US emissions decline is due to switching electricity from coal to natural gas. Obama was bragging about alternating from one fossil fuel to another.

To understand "peak-and-decline" language, we turn to the Kuznets Curve, named for twentieth-century economist Simon Kuznets. The curve should look like an inverted U, where income per capita lies along the X-axis and inequality along the Y-axis. Kuznets hypothesized that as income rose, inequality would first rise and then later decline. Not quite, but that's the theory.

Likewise, a modification of this curve says as income per capita increases (X-axis), so, too, does environmental damage (Y-axis) increase, until a point is crossed at which more income leads to less environmental damage. An environmental Kuznets Curve appears to be true sometimes (reining in CFCs, passing the Clean Air Act). Sustainable development or green growth, as opposed to degrowth, depends on a Kuznets Curve showing up to invert climbing damages with a nice, inverted U pattern. In Haberl et al.'s meta-analysis of 835 decoupling studies, they found that "studies searching for an [environmental Kuznetz Curve] often find no indication for the existence of a turning point" and "there is little support for the inverted U-shape hypothesis."[8]

---

7 I use the updated number of thirty-two declining emissions states, but drivers are from a slightly earlier study when eighteen declined. See Zeke Hausfather, "Absolute Decoupling of Economic Growth," The Breakthrough Institute, April 6, 2021, thebreakthrough.org. See also Corinne Le Quéré et al., "Drivers of Declining $CO_2$ Emissions in 18 Developed Economies," *Nature Climate Change* 9, no. 3 (March 2019): 213–7.

8 Helmut Haberl et al., "A Systematic Review," 29.

Carbon Brief analyzed all reasons US emissions declined around the time of Obama's victory lap. The shift to natural gas accounted for 33 percent of the decline. Wind generation took second place at 19 percent and was closely followed by less electricity usage at 18 percent. Less fuel use and changes in transportation took 12 and 18 percent of the pie, respectively. Increased solar capacity was only 3 percent of the drop.[9] Let us not pretend switching one fossil fuel for another vindicates green growth.

Unlimited growth is impossible, but imagine growth continues for a few centuries while we seek to decouple growth from emissions. Decarbonization is inevitable in the long term (otherwise fossil fuels will be spent or civilization will collapse), but is absolute decoupling possible in the near to medium term? Though there are thirty-two states where emissions are in decline while economies grow, the literature still looks bleak. Beyond those cases, states have achieved only relative decoupling.[10] Proponents of decoupling often seem to imply that absolute decoupling is merely a matter of time. Three rejoinders come to mind. First, for the sake of those who will suffer, I deeply hope the proponents of absolute decoupling are correct, though absolute decoupling seems intuitively limited given the methods used so far.[11] Second, Haberl et al. argue, and I suspect most absolute decoupling optimists would agree, the rate of decoupling is too slow to meet the Paris Agreement.[12] Third, if absolute decoupling can mean switching from coal to gas, perhaps the term is too vague, too open to coalitions between rightful enemies.

Calculating physical limits to growth is at least as old as the 1972 Club of Rome report *Limits to Growth*, in which a team ran simulations and concluded that, if industrialization, pollution, food production, and resource depletion continued unabated, "The most probable result will be a rather sudden and uncontrollable decline in both population and industrial capacity."[13] The report was immediately controversial not only for saying growth must cease but for attempting to model limits.

---

9  Hausfather, "Analysis: Why US Carbon Emissions Have Fallen."

10  Haberl et al., "A Systematic Review," 34.

11  For a thoughtful defense of absolute decoupling, see "Absolute Decoupling of Economic Growth and Emissions in 32 Countries," The Breakthrough Institute, accessed June 24, 2022, thebreakthrough.org.

12  Haberl et al., "A Systematic Review," 33.

13  Donella Meadows et al., *The Limits to Growth* (New York: Universe Books, 1972), 23.

In a report for the European Environmental Bureau, the economist Timothée Parrique argued, "Not only is there no empirical evidence supporting the existence of a decoupling of economic growth from environmental pressure . . . such decoupling appears unlikely to happen in the future."[14] There were seven reasons for skepticism. First, energy becomes more expensive over time. Fossil fuels get more difficult to extract. Renewables will face similar if less pronounced challenges with mining. Second, the Jevons paradox means increased efficiency drives more consumption. Third, solutions to one problem can generate others, as when electric vehicles require more mineral inputs. Fourth, unpredictable impacts will hit the service industry. Fifth, even if recycling capabilities improve, mining will remain crucial. Sixth, technological progress often exacerbates rather than abates destruction. Lastly, decoupling is often a trick of offshore emissions, but carbon dioxide spreads everywhere no matter the country of origin.

Like carbon offsets and cap-and-trade, decoupling can be—not *is by necessity*, but *can be*—a technocratic global game of three-card monte. The marks are duped developed states who take it on faith that growth is good, therefore green growth is necessarily possible. "Policy-makers have to acknowledge the fact that addressing environmental breakdown may require a direct downscaling of economic production and consumption in the wealthiest countries," concluded the EEB report.[15] Sustainable growth is not, to my mind, clearly impossible in the medium term, but what we can say clearly is that there's not yet good evidence it will turn out to be an option at all. Right now, global growth means emissions.

## Physics of the Limits to Growth

"Fairy tales of eternal economic growth" are likely to plague us for a while, tugging at the heartstrings of bourgeois deniers and luring reactionaries with prosperity gospel, so let us exorcise this false orthodoxy. Like we said, infinite growth in a world of finite resources is impossible. To say so conjures a limit imposed by some unknown mineral(s), perhaps lithium or cobalt, but which?

---

14 Timothée Parrique, "Decoupling Debunked," European Environmental Bureau, July 2019, eeb.org, 3.

15 Ibid., 5.

I'll tackle this question by an unfamiliar route. We do know there's a limit to how much energy we can consume. Taking a thought experiment from the physicist Tom Murphy as my starting place to make what, so far as I'm aware, is a novel contribution to Marxist thought, let's take energy consumption as an absolute limit to demonstrate why growth can't continue ad infinitum. This experiment will expose a contradiction between the law of capitalist accumulation (M-C-M') and a concept from physics called the Stefan-Boltzmann law.[16]

Global energy consumption grows 1–2.3 percent per year. So as to not overstate the numbers, I'll model a conservative growth rate of 1.5 percent per year.[17] Energy introduced to a system must be radiated out, and more radiated energy raises a system's temperature, so here's an absurd question: At our present growth rate, how long would it take to turn the surface temperature up to 100°C and boil ourselves?

Set up the problem like so. The sun hits Earth with 174,000 TW. We can meet our 19 TW need by covering half a percent of Earth's land in solar photovoltaic panels operating at 20 percent efficiency. We'd steadily need to cover more land with panels, though, as our wattage needs grew. By 2100, we'd need to cover a whole 1 percent of Earth's land in solar panels. Fine, but growth continues. Problems arise.

In four centuries of 1.5 percent growth, we'd need to cover every square meter of Earth's surface in solar panels. Another century and we'd need to cover the oceans. No rain, plants, or animals could exist. Say humans built a solar farm in space to buy more time, and

---

16 This use of the Stefan-Boltzmann law was inspired by Tom Murphy, who broadly used it to show limits of growth. So far as I'm aware, the use of this concept from a Marxist perspective, pointing to the contradiction between M-C-M' and physics, is original to me. See Tom Murphy, "Exponential Economist Meets Finite Physicist," *Do the Math* (blog), April 10, 2012, dothemath.ucsd.edu; and Murphy, *Energy and Human Ambitions*, 8–14.

17 For comparison, US energy consumption grows at 2.9 percent per year. The global 1.5 percent rate is what we get if energy consumption grows by half by midcentury as forecasts indicate. See "EIA Projects Nearly 50% Increase in World Energy Use by 2050, Led by Growth in Renewables," Energy Information Administration, October 7, 2021, eia.gov; and Hannah Ritchie, Max Roser, and Pablo Rosado, "Energy Production and Consumption," Our World in Data, November 28, 2020, ourworldindata.org.

eventually our descendants, as stewards of growth, decide it best to surround the sun with solar panels and collect its full energy output of $4 \times 10^{26}$ W. In four millennia, that's how much energy we'd need to harvest.

Four millennia sounds like a lot. But in the other direction, writing is older than that. We know what humans were doing then. On that growth rate, we'd need to harvest the sun's full wattage in another four thousand. This would transform the solar system into a cold dead zone, so perhaps we'll sort out nuclear fusion or some yet-to-be-conceived energy source. Either way, if energy consumption were to grow at this pace, the problem for life on Earth would be far worse, far sooner.

Thermodynamic limits to growth work by similar principles to global warming. As energy enters a system, an equal amount must leave the system. If less energy leaves than enters, the system warms up. If more energy leaves than enters, the system cools.

The Stefan-Boltzmann law lets us calculate how hot the surface of an object is in relation to its surface area and radiated energy. I won't detail the minutiae here. If you want to run the numbers yourself using Earth's surface area and energy projections, you can find a Stefan-Boltzmann calculator online. The upshot is that Earth (or any body) has a temperature to which it rises from radiating energy in relation to energy inputs and its surface area. Since it doesn't factor in greenhouse gasses, the Stefan-Boltzmann law says Earth's temperature, given the energy it receives from the sun and radiates back out, should be -18°C. That's the temperature the Stefan-Boltzmann law gives us for an object receiving 123,000 TW (the portion of sunlight striking Earth's surface minus albedo effect) with Earth's surface area of 510 million square kilometers. Earth's actual an average temperature of 15°C is boosted by a 33°C greenhouse gas blanket.

A small surface area radiating a comparatively large wattage, say, a 100 W light bulb, is hot. If Earth's entire surface area emitted a total of 100 W, it would near absolute zero. Energy radiated and the surface area matter for the temperature. If Earth acquired more energy than it receives from the sun, by way of a space-based solar farm or fusion for permanent growth, and assuming there's no feasible mechanism by which all surface energy could be collected, surface temperature would rise with additional wattage.

At 1.5 percent growth rate starting from 19 TW, in 670 years we would be adding 387,000 TW to Earth's system. At that point, the surface would reach 100°C.[18] The previous sentence is not an exaggeration nor a typo, and it doesn't make a difference whether the energy power source is clean. The air would be boiling hot.

Well, maybe subtract five to ten years, depending on how much global warming we add the normal way through carbon dioxide. Or if the entire world caught up to the US growth rate of 2.9 percent, in 314 years we'd reach 10°C warming and, in 350 years, we'd reach 100°C. In seven and a half centuries, Earth's surface would be as hot as the sun's. Again, this is true regardless of how clean the energy source is. In exponential growth scenarios, heat from greenhouse gasses would be a comically small factor.

This is absurd. Humans won't boil themselves. So what objections can we raise about this thought experiment, and what do the real physical limits tell us about the fairy tales of unlimited growth?

The first objection will surely be that if population levels off, so, too, will energy. In *Growth*, Vaclav Smil exhaustively documented exponential growth across many sectors. In one example, the speed of passenger airliners followed an exponential pattern; we went from no planes at all to the Concorde in only decades, but speeds stabilized and are not likely to rise further. If you plotted the rise in airspeed from the Wright brothers' flyer at Kitty Hawk in 1903 to the first flight of the Concorde in 1969, you might conclude all airliners should near the speed of light soon. That's if you plotted growth exponentially without limits, the way some people talk about renewable energy becoming cheaper and cheaper as if there's no floor.

Growth follows a few different patterns, which can be plotted on a simple graph with X-axis and Y-axis. If growth is bounded, not infinite, then it will look either exponential or sigmoidal. In exponential growth, an explosive rise eventually levels out. In S-shaped sigmoidal growth, we see essentially the same pattern of quick rise followed by leveling off, but just before the rise there's also a period that looks nearly flat.[19] The speed of the plane was exponential. Population will be sigmoidal. It took all of human prehistory and history up until the beginning of the nineteenth century to reach a billion people, then fewer than two hundred

---

18  Granted I'm using emissivity for a perfect black body. Actual materials of Earth's surface would change the number by a few years but not significantly.

19  See especially chapter 1 of Smil, *Growth*.

years to reach five billion, and population should stabilize around ten million, give or take.

Energy consumption has risen faster than population increase, though not by much. The global average is 2,300 W per person.[20] Say population stabilized such that there was no more energy growth due to population growth alone. What would happen to economic growth?

If GDP continued to grow with a stable population and capped energy consumption (to prevent boiling ourselves), then the physical economy of goods, all of which take energy, would need to flatline and the service economy pick up the slack. The difference between goods (physical) and services (non-physical) is crucial. If resource consumption is capped, any growth would come from non-physical finance and services ballooning into a larger and larger share of the economy.

In this equally absurd scenario of economic growth with capped resource consumption, the physical economy would continuously shrink as a percentage of the economy. Physical resources would become cheaper without end. If a fixed, scarce resource like energy became infinitely cheap, then we'd have a conflict with the supply and demand principle. Try to imagine our descendants continuing to labor and drive growth while necessities for survival turn infinitely cheap.

There are more contradictions in M-C-M' than Marx knew. Our great fear is that capitalism is too resilient to adapt in response to existential threat. Capitalism may end the way the left dreams. It may not end in time to prevent disaster. On a long enough timeline, though, the law of accumulation will slam into an irresolvable contradiction with physics.

## Climate Economics

I have so far been fully deferential to IPCC analyses, but climate economics has deep uncertainties. Neoclassical economics is an orthodox theology with priests, after all. Economic categories,

---

20   Right now, the biggest offenders in terms of per capita energy consumption are Saudi Arabia and the United States, both around 10,000 W per person. China and India come in at 2,800 W and 600 W, respectively. Europe averages 4,900 W. Murphy, *Energy and Human Ambitions*, 46–7.

like all social sciences, are "forms of expression, manifestations of existence, and frequently but one-sided aspects of this subject," as Marx put it, and they correspond to a definite form of society, in this case capitalist society. We don't have to like discount rates, externalities, or "optimal warming," but we should familiarize ourselves with the master's tools.

Integrated assessment models (IAMs) are used to assess human behavioral impacts on the nature. As the name suggests, they integrate climate and non-climate models. Simple IAMs can be run on spreadsheets gauging monetary costs and benefits for emissions levels, while more complex IAMs account for technological changes, land-use changes, and policy options.[21] IAM modelers are heterogeneous in discipline, not all economists, and IAM research groups submit scenarios to the IPCC for analysis. It's now a large community but wasn't always so. Simple IAMs, including one still cited in IPCC reports, descend genealogically from a modeler who's worked these problems for decades, since long before the implications were clear to economists.

In 2018, William Nordhaus won the fake Nobel Prize that bankers created to legitimate what they do, for his work integrating climate models and economics. While I'm generally skeptical of IAMs, he deserves credit for tackling a difficult problem early. How does one integrate devastating "externalities" into economic models for the future? His pioneering IAM Dynamic Integrated Climate-Economy (DICE) tackles a question of policy: How do you price a commodity that's useful now but destructive later?

Externalities are non-market outcomes of market activities (Nordhaus uses the term "spillover"). "If you poison the groundwater fertilizing your crops, but don't pay the costs created for everyone else, then losses due to poisoned groundwater are not reflected in the price of the farmer's produce," explained Geoff Mann on the problem of externalities. "This misaligns market incentives, over-rewarding destructive behaviour."[22] Remember how fossil fuel subsidies aren't cash payments but rather the externalities for which corporations do not pay? The market underprices things

21 Simon Evans and Zeke Hausfather, "Q&A: How 'Integrated Assessment Models' Are Used to Study Climate Change," Carbon Brief, February 10, 2018, carbonbrief.org.

22 Geoff Mann, review of *Check Your Spillover*, by William D. Nordhaus, *London Review of Books*, February 10, 2022, lrb.co.uk.

constantly, because it treats the world outside of the narrow sliver of the market as an externality.

Nordhaus's model tries to internalize the externalities, most commonly through a carbon tax. Remember the IPCC recommends a tax of $135 to $6,050 per tonne of carbon dioxide by 2030, raising the price to as much as $30,100 per tonne by 2100.[23] But only a few dozen nations implement carbon taxes, and the two with a carbon tax plus dividend paid to citizens, Canada and Switzerland, impose a smaller tax ($30 and $105 per tonne). To correct for spillover and account for the social cost of carbon, DICE recommends a price of about $45, far below what the IPCC recommends.

Skeptical though I may be, we shouldn't discount certain complex IAMs as important research tools. I use an IAM called Global Change Analysis Model (GCAM), one of the primary IAMs used by the research community to study shared socioeconomic pathways. They are tools for conceptualizing futures, and anyone in the IAM research community will tell you projections are not prophecies. Still, if they are not at least somewhat useful as predictions, of what use are they? What the left should notice, I'd argue, is that in shared socioeconomic pathways where emissions are reduced, mitigation is modeled in IAMs through a small number of often dubious methods: Kyoto-style, state-backed limits on emissions, carbon prices, carbon capture and storage, and other policy adjustments in regional groupings rather than state-by-state resolution, and they do not model fundamental changes to political economy. Due to its convenience as a proxy for complex policy, it is often only the carbon price that is toggled to cut emissions in the various scenarios. But IAMs don't care whether the carbon price is a literal carbon tax, a regulatory price set for policy guidance, or a price on exchanges like the EU Emissions Trading System. It's a number in a formula. In the second case of policy guidance, regulators weigh a social cost of carbon for ecological damage wrought by a half tonne of emissions (around fifty dollars at the current US carbon price) against its source in a hundred-dollar barrel of oil, an equation that all too often justifies the leasing of public land for drilling since the carbon price falls well below oil revenues. The regulatory carbon price needn't imply that there's any literal cost in dollars for

---

23 Intergovernmental Panel on Climate Change, *Global Warming of 1.5 °C*, 2018, 152.

emissions, whereas a taxed carbon price would cost in dollars. In their formulae, IAMs needn't distinguish between different carbon price instruments. Furthermore, the regulatory social cost of carbon can be lowered quietly when political winds change, whereas a carbon tax high enough to drive down emissions assumes the capitalist class will tolerate a massive, voluntary loss of revenue.

To save the future, how little should the future be valued? To answer this question Nordhaus uses a discount rate assigning a value to the future relative to today. When someone prefers cheap gasoline today, even if it causes environmental damage later, they discount the future. The number can be quantified. A discount rate of zero means the future and today are worth the same. It's monstrous ethically to weigh the value of future lives differently than today's. Yet we discount constantly.

A discount rate works the same as a return on investment. If I take a loan today at X percent interest, then I have discounted the future (I value receiving money now more than paying a greater sum later). "For the purposes of climate economics," as Mann reframed the problem, "the discount rate is the value, in the present, of the expected benefits and costs of projects to mitigate the possible effects of global climate change."[24] Ecologically unwise but economically lucrative decisions now might boost economies such that we can bear mitigation costs later. Combust all our fossil fuels and GDP might grow enough that cleanup (direct air capture) is a smaller percentage of income later rather than stranding assets now. So there's a theoretically ideal discount rate corresponding to trade-offs of economic growth and warming. Variables have such wide uncertainties our IAMs might as well be guesses.

Using a discount rate of 4.25 percent, Nordhaus suggested optimal warming is 3.5°C.[25] Pause to feel the gravity of this ideal.

Nordhaus drew his discount rate from the market. If the interest rate for a twenty-year bond is in that neighborhood, it's the revealed rate. Nordhaus isn't naive about the problems posed. It doesn't mean 3.5°C is morally right. It's the blind desire of capital.

A caution against straw-manning Nordhaus: we all have an ideal or optimal level of warming implied by our preferred emissions scenario. Say we banned fossil fuels starting today. The loss of

---

24  Geoff Mann, review of *Check Your Spillover.*
25  See chapter 18 of William D. Nordhaus, *The Climate Casino* (New Haven: Yale University Press, 2013).

natural gas for the Haber process would starve at least three billion people in the first year. The sudden loss of eight-tenths of our energy supplied by fossil fuels would trigger a collapse two or three times worse than the Great Depression. Commerce would cease. Governments would topple. Fascists would pledge to turn the wellheads back on and never again countenance clean energy. If this sounds unacceptable, as it does to me, there's additional warming implied in our more humanitarian approach.

If we use Nordhaus's model to plot two lines, (1) the falling costs of mitigation as temperature rises and (2) the rising cost of damages from rising temperature, then the point at which the two meet is 3.5°C. In his model, tipping-point damages cost 9 percent of income at 4°C and skyrocket to 29 percent at 4.5°C, but costs are only 0.5 percent at 3.5°C. We can dispute his ideological priors and inputs (Nordhaus acknowledges deep uncertainties in those inputs), but we'd still all have an ideal warming.

If we are willing to take a bit more warming to prevent unimaginable human costs, then we have an optimal warming. At that point, we're bargaining the level.

Frank Ramsey, the mathematician who pioneered discount rates, suggested any discount rate greater than zero "is ethically indefensible and arises merely from the weakness of the imagination."[26] In his own critique, Mann said IAMs reflect the values of a small minority of wealthy people: "We're asking whether or not this is worth BlackRock's or Citigroup's time and money, and the real rate of return right now would suggest it's not; and then that would say those models are . . . invisibly handing decision-making power over to people who already many of us would mistrust to lead us to the future."[27]

If we weren't running a one-shot existential experiment for the entire biosphere, we could treat neoclassical climate models like a Sokal hoax. So argued the economist Steve Keen's aptly named article "The Appallingly Bad Neoclassical Economics of Climate Change." In his estimation, the problem with these models is a sort of garbage-in, garbage-out formula. "The blasé conclusions they reach," Keen inveighed, "carry far more weight with politicians, obsessed as they are with their countries' GDP growth rates, than

---

26  Ibid.

27  Geoff Mann, "The Climate Colossus," interview by James Butler, *London Review of Books*, February 15, 2022, lrb.co.uk.

the much more alarming ecological warnings in the sections of the Report written by actual scientists."[28] Letting economists grab the headlines of climate reports is a time-tested method for obscuring dire warnings.

Keen focused on two popular methods for economic impact: the enumeration approach and the statistical approach. In the enumeration approach, count up damages and assign a price. The result might turn out ridiculously cheap, such as when the *Fifth Assessment Report* said 2.5°C would mean a loss of income as little as 0.2–2 percent.[29]

Nordhaus posed a similarly low number when, in 1991 (to be fair, it's an old paper), he said climate change's effect on 87 percent of US industry would be negligible and only 3 percent would prove highly sensitive.[30] What kind of industries did Nordhaus say were immune to climate change? He listed hard manual labor like manufacturing and mining as unaffected. "Explicitly, they are saying that if an activity is exposed to the weather, it is vulnerable to climate change, but if it is not, it is 'not really exposed to climate change,'" Keen noted. "They are equating the climate to the weather."[31] Other sectors escaping global warming impacts included real estate, transportation, government services, and insurance. Surely the list would surprise insurance agents today nervously eyeing floods and wildfires.

In the second approach, the statistical or cross-sectional method, economists look at variations in prices across different climates and project how a temperature change would affect price. However, regions with extreme differences from mean temperature also have lower GDP.

The warm climate of Florida (average 22.5°C) and the chilly climate of North Dakota (4.7°C) are far from the US average (12.4°C), while New York (7.4°C) is closer. It turns out that New York's per capita income is higher than North Dakota's. Wow.

In this method, heat would push both North Dakota and New York closer to the mean and benefit both. Warming will benefit colder regions in limited ways, but here's a headline finding from

28   Steve Keen, "The Appallingly Bad Neoclassical Economics of Climate Change," *Globalizations* 18, no. 7 (October 3, 2021): 1149–77.

29   Intergovernmental Panel on Climate Change, *Climate Change 2014: Synthesis Report*, 2014, 73.

30   William Nordhaus, "To Slow or Not to Slow," *The Economic Journal* 101, no. 407 (July 1991): 920–37.

31   Keen, "The Appallingly Bad Neoclassical Economics," 5.

one such study: "$97 to $185 billion of benefits, with an average of $145 billion of benefits a year."[32] The full findings are truly incredible and claim agriculture will reap windfall profits in excess of a hundred billion while other fields warn of famine.[33]

Asking experts to read our futures is the desire for a subject-supposed-to-know. Hegemonic neoclassical methods suck the oxygen out of the room. To his credit, Nordhaus knows his model paints a rosier picture than climatologists foresee. In one of his surveys, natural scientists estimated future damages twenty to thirty times higher than economists. One scientist mocked the folly of such guesswork:

> I must tell you that I marvel that economists are willing to make quantitative estimates of economic consequences of climate change where the only measures available are estimates of global surface average increases in temperature. As [one] who has spent his career worrying about the vagaries of the dynamics of the atmosphere, I marvel that they can translate a single global number, an extremely poor surrogate for a description of the climatic conditions, into quantitative estimates of impacts of global economic conditions.[34]

A drop in emissions doesn't have to mean a drop in productive output, but our current mode of production and distribution of resources suggests it would right now. Over the last half century, energy and global GDP had a correlation coefficient of 0.997.[35] Even now, the correlation coefficient for carbon dioxide concentration and GDP is 0.958.[36] You can see the outlier year of pandemic lockdowns. We haven't much evidence the marriages will divorce.

32   Robert Mendelsohn, Michael Schlesinger, and Larry Williams, "Comparing Impacts Across Climate Models," *Integrated Assessment* 1 (2000): 37–48.

33   Ibid., 42.

34   William D. Nordhaus, "Expert Opinion on Climatic Change," *American Scientist* 82, no. 1 (1994): 45–51, 51.

35   Keen, "The Appallingly Bad Neoclassical Economics," 14–15.

36   Figure 8.1 on the following page sourced from: C. D. Keeling et al., "Primary Mauna Loa $CO_2$ Record 1958–Present," Scripps $CO_2$ Program (San Diego: Scripps Institution of Oceanography, 2022); and "GDP Per Capita (Constant 2015 US$)," The World Bank, accessed February 23, 2023, data.worldbank.org.

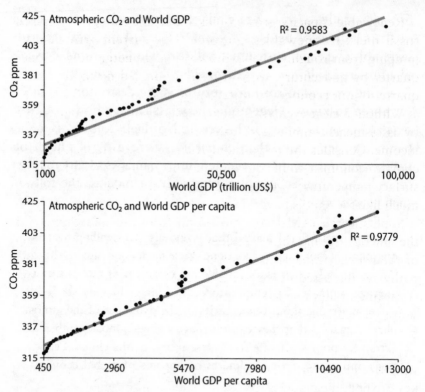

Figure 8.1. Carbon Dioxide and GDP

Fossil fuels deliver eight-tenths of global energy. If we instantaneously cut fossil fuels today, GDP should plummet in tandem. Renewable electricity generation worldwide recently jumped from a paltry 3 percent per year growth rate to 7 percent by 2020.[37] But total energy use grows too. The Energy Information Administration projects renewables will still make up only a quarter of world energy consumption by midcentury.[38] EIA projections aren't predictive forecasts. In fact, the organization expects projections to miss due to policy changes. What a projection implies is the current course. Likewise, the International Energy Agency scenarios often underestimate renewables growth, so take this with caution, but in its most recent forecast after the massive pandemic year boost in renewables, near-term renewable growth was only around half of what it needed to be to get to net zero by 2050.

37 "Renewables—Fuels & Technologies," International Energy Agency, 2021, iea.org.

38 "International Energy Outlook," Energy Information Administration, October 6, 2021, eia.gov.

Renewable capacity grows quickly but not enough to replace fossil fuels. If renewables continue their current growth rate, jumping from roughly one-sixth of primary energy now to one-quarter by midcentury, we will have decoupled no more than a quarter of our economy from carbon dioxide.

Without a clear-eyed assessment, our policymakers might be led by assessments claiming 2.5°C shaves off as little as 0.2 percent of income. Consider the response of Richard Tol, a frequently cited climate economist, to the question of whether a 10°C rise in global surface temperatures would be manageable: "We'd move indoors, much like the Saudis have."[39]

Sophisticated IAMs have moved on from the spreadsheets of the nineties. Taking the example mentioned above, GCAM is one of the primary models used in the IPCC's shared socioeconomic pathways discussed in chapter 4. GCAM is a multi-module IAM consisting of files toggled to input populations, taxes, policy solutions, greenhouse gas forcings, and dozens of other options. These files are read in and generate an output file showing consequences of everything from radiative forcing to how many hectares of land will be required for agriculture across thirty-two global regions. A key driver in the model is GDP, calculated by projecting population and labor productivity in five-year intervals out to 2100. Is it possible to do that in a meaningful way? Slight changes to any input change all other outputs.

I feel a hesitant respect for what the modeling community tries to accomplish. Limitations are everywhere and are well known in the modeling community. IAMs often use a Solow-Swan exogenous growth approach and calculate output using a Cobb-Douglas function. Marxian and even Keynesian functions are absent in this neoclassical silo. IAMs have a supply-side bias wherein lowest energy prices determine which energy sources are used, though we have ample evidence the lowest price does not determine energy sources. Most surprising, IAMs do not count costs of climate damage.[40]

IAMs are also highly technical and exclude anyone without expertise, which can mislead policymakers and the public. Take

---

39  Richard Tol as quoted in Keen, "The Appallingly Bad Neoclassical Economics," 17.

40  Rob Dellink and et al., "Long-Term Economic Growth Projections in the Shared Socioeconomic Pathways," *Global Environmental Change* 42 (January 2017): 200–14.

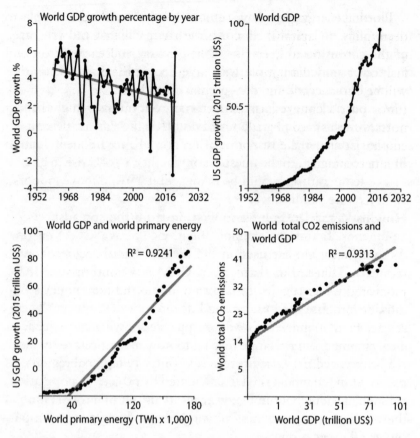

Figure 8.2. Relationships and Challenges to Decoupling

for example SSP1-1.9, which an IPCC Summary for Policymakers presents as a pathway for limiting warming to 1.5°C with no overshoot. Does the public understand that this scenario implies a more or less overnight transition to global social democracy plus trillions spent on carbon capture? Given all their flaws, climatologist Kevin Anderson said IAMs "have been as damaging to the agenda of cutting emissions as Exxon was in misleading the public about climate science." Climatologist Johan Rockström called IAM scenarios "academic gymnastics that have nothing to do with reality."[41]

41   Kevin Anderson, "IPCC's Conservative Nature Masks True Scale of Action Needed to Avert Catastrophic Climate Change," The Conversation, March 24, 2023, theconversation.com, and Kevin Anderson and Jessica Jewell, "Debating the Bedrock of Climate-Change Mitigation Scenarios," Nature 573, no. 7774 (September 2019): 348–49.

Plotting energy growth and emissions, we see other challenges to decoupling.[42] Capitalist climate governance will take full advantage of the confusion to keep the profits rolling in. Reactionaries will lash out at any disruption. Real advances in the energy transition will be bureaucratic and dull, not flashy. Rather than hope to sneak funds past a congress or parliament in omnibus bills, we must normalize the need for trillions of dollars in public investment and emphasize that the costs of inaction are far more than a fraction of a percentage point.

## How to Waste What Time We Have Left

As room left in the carbon budget dwindles, cynical predators with money and libertarian brain, along with petty bourgeois aspirants, pioneered new ways to flush money down the toilet and carbon dioxide into the air. In recent years, thousands of digital cryptocurrencies flooded markets. The numbers in this section were current as of summer 2022; though they will have changed by the time you read this (not least due to market crashes while this book went to press), I leave them unaltered as a testament to cynicism and financial denial. Cryptocurrencies are speculative gimmicks preying on the public, but due to the vast amounts of energy consumed by server farms mining for new coins, if they gain widespread use, they'll cook us.

If Bitcoin was adopted at the fastest rate of inclusion for new technologies, a 2018 study showed it could bust the carbon budget for 2°C in eleven years (twenty-two years for medium rate adoption).[43] As I write this four years later, Bitcoin's carbon footprint has fully doubled per accounting by Alex de Vries. Bitcoin consumes electricity north of 200 TWh per year. In those units, global energy

---

42    Figure 8.2 GDP numbers from "GDP Growth (Annual %)," The World Bank, accessed February 23, 2023, worldbank.org; and "GDP (Current US$)," The World Bank, accessed February 23, 2023, worldbank. org. Energy numbers from Hannah Ritchie and Max Roser, "Energy Mix," Our World in Data, 2022, ourworldindata.org; and Hannah Ritchie, Max Roser, and Pablo Rosado, "$CO_2$ Emissions," Our World in Data, 2020, ourworldindata.org.

43    Camilo Mora et al., "Bitcoin Emissions Alone Could Push Global Warming Above 2°C," *Nature Climate Change* 8, no. 11 (November 2018): 931–3.

consumption comes to around 165,000 TWh per year. Bitcoin is wasting 0.8 percent of global electricity.[44] Energy consumption has been rising steeply though, so by the time you read this, it could be in excess of a hundredth of global electricity.

It's common to hear Bitcoin uses more electricity than this or that country, which clouds the size of the problem. It's not just a few countries; Bitcoin uses more electricity than all but twenty-two countries.[45] Its total carbon footprint is one-quarter of a percent of all emissions (114 $MtCO_2$ out of 41.4 $GtCO_2$), and each Bitcoin transaction generates a carbon footprint of 1,200 $kgCO_2$ or about as much as an American emits in a month. Other cryptocurrencies fare better but aren't good. Even after a market-wide cryptocurrency plummet in the final days writing this book, a single Ethereum transaction is worth three days of an average American's emissions, and Dogecoin is worth seven.[46]

Needless to say, this is controversial. When cryptocurrency promoters wrote in the conservative rag *Washington Examiner* an article titled "No, Bitcoin Is Not Destroying the World," a group of self-styled green Republicans argued that Bitcoin consumed only 40 percent as much as the banks and gold mines, as if this were a good feature. They underscored how three-quarters of coin miners use at least some renewable energy, which is true when plugging in to almost any grid in the world. A never-ending supply of bad faith justifications guards the conscience from what is plain. Marketing it as a get-rich pyramid scheme lays bare the nihilism of capital. We are running out of time, and what is finance doing?

Declining population growth will probably relieve pressure on resource-consumption growth. This will not save capitalism long term, as we have seen, since it will only slow the rate at which we either cook ourselves with too much energy or enter a twilight zone of infinitely cheap resources, but it will buy time. There are positive signs in this direction. Primary energy use grew faster in the first half of the twentieth century than it has since. However, even

---

44 Alex de Vries, "Bitcoin Energy Consumption Index," Digiconomist, accessed May 3, 2022, digiconomist.net; Hannah Ritchie and Max Roser, "Electricity Mix," Our World in Data, 2020, ourworldindata.org; and Ritchie, Roser, and Rosado, "Energy Production and Consumption."

45 Colm Hebblethwaite, "Bitcoin Mining Electricity Usage Higher Than 159 Countries," The Block, November 24, 2017, blockchain technology-news.com.

46 De Vries, "Bitcoin Energy Consumption Index."

as growth of energy slows, the total amount of energy consumed is still rising. One way or another, the motion will grind down on real biophysical limits.

"There is no possibility of reconciling the preservation of a well-functioning biosphere with the standard economic mantra that is akin to positing a *perpetuum mobile* machine," Smil said of the fantasy of unlimited growth without problems of sustainability. Economists "maintain a monopoly on supplying their physically impossible narratives of continuing growth that guide decisions made by national governments and companies."[47] Against this type of unabated growth, or unabated growth plus carbon capture and some renewables, there are competing proposals for either green/sustainable growth or steady-state growth and degrowth.

Capitalism constantly underprices commodities, because it doesn't pay for environmental or social damage. A "social cost of carbon" or a "risk cost of carbon" could fix this, a sort of modified and targeted carbon tax. A social cost of carbon accounts for the "negative externality" created by anthropogenic greenhouse gasses. The social cost would account for the loss of economic welfare from each tonne of emissions, while the risk cost of carbon would count up roughly what an insurance adjuster might assess as damages from climate change. If the expense of future damage can be quantified, it can be shown that paying for mitigation today is cheaper than paying for damage tomorrow. In limited instances, we now use a low-ball social cost of carbon for regulation (the Obama administration put the social cost of carbon at fifty dollars; the Trump administration dropped it to as little as one dollar, and Biden returned it to the prior level).

Combining the two concepts could give us a new type of currency instrument (some dub it a "carbon coin") by which central banks might incentivize decarbonization. Just as we pumped up liquidity in the 2008 financial crisis and the 2020 pandemic, Delton Chen's carbon coin opens an option he calls "carbon quantitative easing."[48] If fossil fuels reserves are worth a couple of hundred trillion dollars,

---

47   Smil, *Growth*, 507.

48   Delton B Chen, Guglielmo Zappalà, and Joel van der Beek, "Carbon Quantitative Easing: Scalable Climate Finance for Managing Systemic Risk," Global Carbon Reward, October 15, 2018. See also Delton Chen, Joel van der Beek, and Jonathan Cloud, "Hypothesis for a Risk Cost of Carbon," in *Understanding Risks and Uncertainties in Energy and Climate Policy*, ed. H. Doukas, A. Flamos, and J. Lieu (Springer, 2018).

or $2.4 trillion in total untapped resources, we can do worse than pay ourselves to not burn all those fuels.

Perhaps there are no green paths. Sustainable growth models like a Green New Deal or carbon quantitative easing are essentially a type of green Keynesianism, which might be better than what we have now but is critically flawed in the long term. "Keynesianism in any one nation assumes and requires a sovereign state monopolizing both the legitimate use of violence and the legitimate allocation of resources within its territory," wrote Joel Wainwright and Geoff Mann in their excellent book *Climate Leviathan*. "But planetary warming exposes the territorial national-state as insufficient to address the crisis."[49] Liberalism sees the state and market oscillating between zero-sum competition and partnership (they will say state intervention "distorts" the market). But the market cannot solve this problem, and no climate leviathan or planetary sovereign has yet appeared to manage the transition. Instead, we have barely a hope of Chinese leadership in the face of lumbering US sabotage of COP agreements and flailing EU leadership. Green Keynesianism could soften the blow only by stitching together ad hoc climate policy nation by nation. There may be no good answer.

All proposals for sustainable growth depend upon decoupling. In Haberl's meta-analysis of decoupling studies, degrowth gets far less attention than green growth. But degrowth is poorly defined, and sustainable growth doesn't have much evidence in its corner either. Parrique acknowledged the problem of so many vague definitions of degrowth since the term *décroissance* gained popularity in the early 2000s. In his working definition, degrowth is "a reduction in the importance of economistic thoughts and practices in social life."[50]

Jason Hickel defined degrowth as "a planned downscaling of energy and resource use to bring the economy back into balance with the living world in a safe, just and equitable way."[51] We can wind down private jets, arms, SUVs, and fossil fuels while growing necessary sectors like health care, green energy, social services, and regenerative agriculture. With a jobs guarantee and a fair distribution of income, an economy producing necessary goods would

49  Joel Wainwright and Geoff Mann, *Climate Leviathan* (London: Verso, 2020), 125.

50  Timothée Parrique, "The Political Economy of Degrowth" (PhD diss., Université Clermont Auvergne, 2019), 34.

51  See introduction in Jason Hickel, *Less Is More* (London: William Heinemann, 2020).

not threaten household incomes, since prices are the obverse of incomes. Hickel is certainly right, though the demand for a fair distribution of income does a lot of work in this proposal. Without surplus accumulation at the top, the system wouldn't be capitalist. We can look as well to Matthias Schmelzer, Andrea Vetter, and Aaron Vansintjan, who define degrowth as:

> The democratic transition to a society that—in order to enable global ecological justice—is based on a much smaller throughput of energy and resources, that deepens democracy and guarantees a good life and social justice for all, and that does not depend on continuous expansion.[52]

Degrowth is not austerity or "romantic luddism," they argue, but a serious acknowledgment of ecological limits in a mode of production hostile to human flourishing.

In Smil's analysis, degrowth is "inelegant and inaccurate newspeak" that displays the predicament of "using regress as a qualifier of civilizational achievement, after a long-lasting addiction to progress." It seems unrealistic in calling for deliberately declining levels. But Smil is not entirely skeptical, because what degrowth hints at is "an irreconcilable conflict or, more accurately, a challenge for which we have yet to find an effective solution (assuming one exists)."[53] Again, maybe no effective solution exists.

Some sort of degrowth, steady-state, or planned economy model might become necessary if sustainable growth proves impossible. The IPCC noted degrowth as an option in its latest report. Various alternatives exist with varying optimism in markets. Doing justice to this topic would take a book, not a chapter, and that book would delve much deeper into the eco-socialist tradition than the gestures I've made throughout this book. To my mind, what's clear is the need for a socialism conscious of ecological limits to growth and conscious of justice for human and non-human life alike. The ideal must be socialist, because workers should own the means of production, because production should be organized for flourishing not surplus accumulation, and because all versions of capitalism depend on exploitation that, in practice, is racist. The ideal must

---

52  Matthias Schmelzer, Andrea Vetter, and Aaron Vansintjan, *The Future Is Degrowth* (London: Verso, 2022), 4.

53  Smil, *Growth*, 510–11.

respect ecological limits because they are physical and inflexible. The ideal must respect non-human life if we seek justice. More strongly, "growth" mystifies and justifies violence. We must challenge the obtuse, reflexive treatment of growth as necessarily good. But whatever its merits or moral high ground, under the capitalist law of accumulation, who has agency to implement degrowth?

What is certain is that capitalism is failing the biosphere and all but the wealthiest of its inhabitants. Neoclassical priests of a particular mode of production write scathing apologetics against other faiths, but their doctrines seek neither human flourishing nor ecological health. What is uncertain is whether there is even a possibility of sustainable growth or decarbonization at the pace needed within a system of pseudo-democracies masking oligarchies. Most certainly liberal capitalism proves itself resilient time and time again. As the manager of capitalist climate governance, its fantasies deserve special scrutiny. Liberal capitalism gives and takes away, but under stress, it gives us just enough treats to keep us happy. Resilience in pursuit of emptiness raises a question of desire: What does the liberal want?

# PART IV

## *Management*

# 9

# What Does the Liberal Want?

*Liberalism, with its contradictions and compromises, existed . . .
only in that short interim period in which it was possible to answer
the question "Christ or Barabbas?" with a proposal to adjourn or
appoint a commission of investigation.*

—Carl Schmitt

*The fundamental problem of political philosophy is still . . . "Why
do men fight for their servitude as stubbornly as though it were their
salvation?"*

—Gilles Deleuze and Félix Guattari

Young activists from the Sunrise Movement petition California
Democratic senator Dianne Feinstein in February 2019. "We are
trying to ask you to vote yes on the Green New Deal," says one
child.

"OK, I'll tell you what. We have our own Green New Deal," the
senator blithely lies.

Indicating a new report on warming at 1.5°C, the child continues,
"Some scientists have said that we have twelve years to turn this
around."

"Well, it's not going to get turned around in ten years," the
senator shrugs. "What we can do is put ourselves—"

Another interjects, "Senator, if this doesn't get turned around in
ten years, you're looking at the faces of the people who are going
to be living with these consequences."

The first child starts again earnestly. "The government is supposed to be for the people, and by the people, and all for the
people."

Feinstein grows impatient. "You know what's interesting about this group? Is I've been doing this for thirty years. I know what I'm doing."

This callous dismissal will grab headlines. The senator continues, "You come in here, and you say it has to be my way or the highway. I don't respond to that." She crosses her arms and rocks forward over the children to whom she talks down. "I've gotten elected, I just ran. I was elected by almost a million-vote plurality," she insists. "And I know what I'm doing. So, you know, maybe people should listen a little bit."

Another activist speaks up. "I hear what you're saying, but we're the people who voted you [sic]. You're supposed to listen to us, that's your job."

"How old are you?" asks the senator. Upon hearing the activist is only sixteen, Feinstein replies, "Well, you didn't vote for me."

A child castigates the senator, "It doesn't matter. We're the ones who are going to be impacted!"

"I understand that. I have seven grandchildren. I understand it very well."

Another reminds her of what the IPCC's report has just made widely known. "We're asking you to be brave and do this for us and for your grandchildren."

Feinstein remains unfazed. "I'm trying to do the best I can, which was to write a responsible resolution."

The activist persists, "Any plan that doesn't take bold, transformative action is not going to be what we need."

Feinstein has had enough. "Well, you know better than I do," she mocks them. "So I think one day you should run for the Senate, and then you do it your way."

Once more a child tries again in good faith, "By that time there's going to be a big problem—"

The senator cuts her off, "I just won a big election."

## The Fool and the Knave

Reading horoscope columns in the *Los Angeles Times* in 1953, Theodor Adorno found a signal on repeat: everything will be fine.[1] Astrology, in his estimation, wasn't irrational *tout court*

---

1   Theodor Adorno, *The Stars Down to Earth*, ed. Stephen Crook (London: Routledge, 2007), 55.

but, instead, a pseudo-rationality for the overly self-involved, a regression in consciousness, and a "rational self-preservation 'run amuck.'"[2] God is dead. So we found in the heavens a mythology even more untrue than abandoned theologies. Everything will be fine.

As a symptom, astrology indicates a regression in consciousness. The occult mitigates anxieties, taps narcissism, and vaguely gestures at someone or something in charge. That would be nice.

What does the liberal want? The conservatives get the ire, rightly so, and there is no doubt that conservative agendas pose the greater intentional threat to life on Earth. It's difficult to imagine a more dangerous organization than the GOP. But if conservatives holler, "Drill, baby, drill!" as Sarah Palin's slogan jeered, what do we make of Joe Biden saying, "I am not banning fracking" (after which Democratic opposition to fracking dropped sixteen points)?[3]

In his seminar, Jacques Lacan posed a question about interpreting convoluted desires: fool or knave? When lies spill from our opponents, do they foolishly believe the lie, or do they know? Why do liberal politicos, those professional believers in nothing, so consistently fall for the obvious lies of liars?

Notwithstanding the American tendency to misclassify liberalism as a left-leaning position, liberalism is theoretically the center of the right-left spectrum. The right clings to hierarchy (capitalist, racist, misogynist, theocratic), and the left works toward egalitarianism. Theoretically centrist, yes, but liberalism's support for individual liberties (speech, education, expression, religion, formal legal equality) plus its unwavering support for capitalist production and class hierarchies mean its governance veers center-right. Liberal leaders consistently align with the right against the left, because they agree with the right on class oppression. For example, the liberal hates to discover anyone is denied housing on account of race or sexual orientation, yet they feel some people ought to be wage laborers while others deserve to extract rents as landlords. Liberalism's contradictions are irresolvable because the worker-owner distinction requires liberalism to pick a side. The liberal's deep conviction is some hierarchies are natural or necessary, and they confess those convictions indirectly so as not to hurt their own feelings or damage

---

2  Ibid., 34.

3  Julian Brave NoiseCat and Danielle Deiseroth, "Earth to Biden: Stop Talking About Fracking," *The Nation*, October 21, 2020, thenation.com.

their self-righteousness. Did Feinstein tell those kids to fuck off because rumors of her dementia were already true, or did she express the standard Democratic policy of kicking the can down the road? When you can't tell, what does the lack of distinction suggest?

"The 'fool' is an innocent, a simpleton," Lacan began his riff, "but truths issue from his mouth that are not simply tolerated but adopted, by virtue of the fact that this 'fool' is sometimes clothed in the insignia of the jester." Indeed, much of the humiliating nonsense we tolerate gets justified on the grounds that our policymakers are too stupid to know better. Lacan argued the progressive evokes a certain foolishness or direct authenticity. It isn't hard to figure out precisely why activists cry out for health care or a habitable biosphere. This needn't only be true of the left. Even among liberal centrists, one finds a fair number of causes in and of themselves worthy of dedicating one's life to.

Not so on the right. An "unmitigated scoundrel," the right-wing intellectual plays the part of a knave. "In other words," Lacan continued, "he doesn't retreat from the consequences of what is called realism; that is, when required, he admits he's a crook." He's a pragmatist speaking harsh truths, but when pressed he'll admit he's a villain and smile, reminding you his side is winning regardless. The right-wing public figure isn't naive. You can't stop him by calling out hypocrisy. He knows. It's his job.

Reactionaries don't see the violent clowns telling them what to believe on television as liars. But ever since Plato first identified this phenomenon in *The Republic*, if a shameless brute robs and dupes the people who invested their trust in him, they'll admire him more. If a genuine fool who believes the dogma gains office, establishment conservatives recognize him as a pest to manage. Once you know this dynamic—fool and the knave as well as the collective desire of fools for a knave—you see it everywhere.

Bourgeois parties find no trouble taking advantage of these confusions in constituencies. Capitalist climate governance and the capitalist denial ISA play off one another. The liberal policymaker depends on their constituents' desire to be blameless and smart, to trust in a manager-supposed-to-know. Thus, "believe the science" is a mantra for social media, marches, and yard signs among those who vote, vote, vote for centrists who promise never to ban fracking.

When Adorno analyzed horoscope columns and found their basic message to be "everything will be fine," he noticed something else of use. People enjoy superstitions when they look official, when they appear systematized, normalized, or even institutionalized. People take the charts half-seriously and don't wish to know how it "really works" or even if it works. "The type of people we are concerned with take astrology for granted, much like psychiatry, symphony concerts or political parties," Adorno said. "They accept it because it *exists*, without much reflection, provided only that their own psychological demands somehow correspond to the offer. They are hardly interested in the justification of the system."[4] Interrupt a fantasy someone is in the process of enjoying (their horoscopes or Teslas), and the fantasizer feels annoyed rather than thankful, especially if the fantasy was negating guilt.

## The Green Dream or Whatever

Not long after Democrats retook the House of Representatives in the 2018 midterm, soon-to-be Speaker Nancy Pelosi, riding a wave of anti-Trump enthusiasm, announced in December her plans for a Select Committee on the Climate Crisis.[5] This was a resurrection of the Select Committee on Energy Independence and Global Warming, which lived briefly from 2007 until Republicans killed it in 2010. Hardly a big loss, the committee wasn't a player in the 2009 Waxman-Markey bill, the lost battle over cap-and-trade legislation. Likewise, Pelosi's new committee sailed into obscurity without a mandate to write a climate bill while activists demanded a new push.

A week after the election, 150 young protesters with the Sunrise Movement set up outside Pelosi's office. Yellow signs read, "GREEN JOBS FOR ALL" and "WHAT IS YOUR PLAN?" Freshman representative Alexandria Ocasio-Cortez joined. They had a nascent idea imported from a UK plan developed a decade prior, a framework built around both environmental and social crises. A Green New Deal.

After Ocasio-Cortez, along with Representative Rashida Tlaib and allies, introduced the resolution on a Green New Deal upon

---

4  Adorno, *The Stars Down to Earth*, 36.
5  Ryan Grim and Briahna Gray, "Alexandria Ocasio-Cortez Joins Environmental Activists in Protest at Democratic Leader Nancy Pelosi's Office," *The Intercept*, November 13, 2018, theintercept.com.

taking office, the legislation turned into a punching bag for the right. Conservatives concocted a Byzantine simulacrum of a bill only fourteen pages long. President Trump represented the right-wing imaginary when he tweeted the Green New Deal would "permanently eliminate all Planes, Cars, Cows, Oil, Gas & the Military."

Speaker Pelosi echoed the right's hostility to a Green New Deal and feigned ignorance, as if she, too, couldn't be bothered to read a few pages. "It will be one of several or maybe many suggestions that we receive," Pelosi blew it off. "The green dream or whatever they call it, nobody knows what it is, but they're for it right?"[6]

After retaking the White House two years later, Democrats redoubled efforts to bargain down their position. "I was convinced, and I remain convinced, no government is going to solve this problem," the new climate envoy John Kerry confessed to the bankers. "The solution is going to come from the private sector, and what government needs to do is create the framework within which the private sector can do what it does best, which is allocate capital and innovate." Treat it like war, he implored.

Emphasizing jobs and new technologies, prominent watchwords for the public-private partnership faith, Kerry spoke inspiringly of a "race to the new technology" and "opportunities that are going to create enormous wealth for those that are venturesome and go out and chase those gold pots." The secretary of the treasury and former federal reserve chair Janet Yellen echoed his estimation that federal funds won't solve the problem, saying, "Private capital will need to fill most of that gap."

When he met with Chinese counterparts, Kerry said the situation was "beyond catastrophic" but "you don't have to give up a quality of life."[7] Catastrophic but not serious. He bet the future on nonexistent tools. "I am told by scientists that 50 percent of the reductions we have to make to get to net zero are going to come from technologies that we don't yet have." It sounds as if he heard someone explain pathways for net zero but confused an option with an inevitability, which modelers will tell you is the danger of showing policymakers these scenarios.

---

6   Heather Caygle, Sarah Ferris, and John Bresnahan, "'Too Hot to Handle': Pelosi Predicts GOP Won't Trigger Another Shutdown," *Politico*, February 7, 2019, politico.com.

7   "World's Largest Iceberg Melts Away After Three Years," YouTube video, posted by Sky News, 2021.

Lacan joked that in every age there are figures akin to alchemists pleading with the powers that be: "Give us money; you don't realize that if you gave us a little money, we would be able to put all kinds of machines, gadgets and contraptions at your service."[8] Often, the result is wasted time or, occasionally, Dr. Oppenheimer transforming knowledge into death, the destroyer of worlds. "Science is animated by some mysterious desire," Lacan warned, "but it doesn't know, any more than anything in the unconscious itself, what that desire means. The future will reveal it to us."[9] Imagine, a few decades from now, the moral rot of technological utopians who told policymakers the situation was dire but would be solved with mitigation gadgets and venture capital.

Hedging on future technologies or corporations is a symptom. We can't see a way out right now. An old, critical flaw in human group behavior: the fantasy that there's a subject-supposed-to-know with solutions mapped out.

Today, everyone wants a subject-supposed-to-know who can resolve obviously critical problems, and for liberals "the science" gets anthropomorphized as an agency or deity with solutions. In his final State of the Union address, President Obama said, "Look, if anybody still wants to dispute the science around climate change, have at it. You will be pretty lonely." Staffers running his social media posted the "you'll be pretty lonely" quote and added a reminder that 97 percent of scientists agree.

For the 2010s and beyond, it's been a strangely fixed number. Ninety-seven percent first appeared in a 2009 University of Illinois survey of climatologists, and the National Academy of Sciences replicated it a year later with a large survey of nearly fourteen hundred climatologists when 97 to 98 percent attributed warming to anthropogenic emissions. Four years later, a meta-analysis of more than ten thousand peer-reviewed articles found 97.1 percent agreement.[10] As the science developed, the number grew. By 2016, out of seventy thousand authors of journal articles only four rejected the

8 Jacques Lacan, *The Ethics of Psychoanalysis: The Seminar of Jacques Lacan, Book VII,* ed. Jacques-Alain Miller, trans. Dennis Porter (New York: W. W. Norton, 1997), 324–5.

9 Ibid., 325.

10 John Cook et al., "Quantifying the Consensus on Anthropogenic Global Warming in the Scientific Literature," *Environmental Research Letters* 8, no. 2 (May 15, 2013).

consensus, a 0.0058 percent denial rate.[11] But as evidence mounts, the cited number never moves, as if its consensus shouldn't be so unanimous as to appear rigged.

The ink wasn't yet dry on the Paris Agreement when Obama lifted the oil export ban. He spoke the right words yet proudly took credit for a fossil fuel boom. Adorno said the proliferation of data meant heaps of facts replaced intellectual penetration.[12] People's wish for a higher power and their wish to reinforce their own dependence led to faith in what Adorno called "prophets of deceit" who promise everything will be fine.[13]

Remember Hansen's critique of the IPCC's scenarios to meet Paris goals. There was no way to meet 1.5°C, so they devised options for stripping carbon dioxide from the air with technologies that barely exist. When I read climatologists working on mitigation with carbon capture, they are perfectly clear about the scenarios being unlikely long shots. It is in policymakers and the public that the fantasy persists uninterrupted.

## Ambitions Interrupted

How would self-professed science believers respond to climate policy actually operating at the level required? Two recent experiments carried out in the United States and UK simultaneously show the general contours of liberal climate denial, of how commitments to capital accumulation and class inequality derail ambitious climate action.

On one side of the Atlantic, out of a clown car of several dozen candidates, the top choices for the 2020 Democratic US presidential campaign were always Joe Biden or Bernie Sanders, with perhaps five others as semi-serious contenders. A democratic socialist, Sanders occupied the center-left of an otherwise uninspiring field and though fighting an uphill battle against a media blackout, gained wide support from key Democratic demographics including young voters, women, people of color, and workers. After Sanders won majorities of votes in the first three primaries or caucuses and

11   James L. Powell, "Climate Scientists Virtually Unanimous: Anthropogenic Global Warming Is True," *Bulletin of Science, Technology & Society* 35, nos. 5–6 (March 28, 2016) 121–4.

12   Adorno, *The Stars Down to Earth*, 119.

13   Ibid., 114.

came second in the fourth, the party intervened and instructed two center-right candidates to drop out so their supporters would flock to Biden. One centrist stayed in the race when she received an influx of big money to split the remaining progressives and avoid consolidation around Sanders. The attack was effective.

For our purposes, the primary raised questions about the importance of climate change to voters who imagine it important. Running on a $16 trillion climate proposal projected to save $70 trillion by end of century, Sanders proposed the most ambitious climate plan ever from a US presidential candidate. The Green New Deal listed fully renewable electricity and transportation no later than 2030, millions of new jobs, full decarbonization of the economy by mid-century, investments in sustainable agriculture, billions for frontline communities, a just transition for fossil fuel workers, investments in public transportation, infrastructure, and high-speed internet, and hundreds of billions in adaptation aid to developing nations.[14]

On the other side of the Atlantic, Jeremy Corbyn and Labour pushed similarly ambitious climate legislation. Labour's manifesto for a Green New Deal included 90 percent electricity generation from low-carbon sources by 2030, a climate apprenticeship program to help businesses develop clean tech skillsets, massive investments in wind turbines and solar panels ("enough solar panels to cover twenty-two thousand football pitches"), an affordable and sustainable integrated public transport system, upgrading energy efficiency for homes, 3 percent GDP diverted to research and development to usher in a "Green Industrial Revolution," an immediate and permanent ban on fracking, and delisting companies from the stock exchange if they fail to contribute to tackling climate change.[15]

Voters in both nations, many of whom claimed to care deeply about global warming and even placed it at the very top of their concerns, opted against these ambitious plans. Perhaps, when faced with an existential crisis, people misjudge threats. Certainly, the ruling classes prefer conservatism, even risking fascism, when the alternative is social democracy or socialism. I've shown my cautious optimism and concerns over a green Keynesianism already, but

---

14 "The Green New Deal," Bernie Sanders Official Website, accessed May 4, 2022, berniesanders.com.

15 "A Green Industrial Revolution," The Labour Party, accessed May 4, 2022, labour.org.uk.

what we saw with Sanders and Corbyn is that, actually, the liberal bourgeoisie rejects any minimal mitigation if it detects egalitarianism in the program. It can manipulate and distort ideology by the manufactured consent of media, such as in portrayals of Corbyn and Sanders as sexists, racists, and anti-Semites.

Accusations of anti-Semitism hit Labour for years leading up to the 2019 election. A few dozen individuals within Labour clearly were guilty of instances ranging from offensive social media posts to outright Holocaust denial.[16] In a case of indefensible misjudgment, Corbyn defended a London mural using the anti-Semitic trope of Jewish bankers taking advantage of the poor. Investigations never found evidence of widespread anti-Semitism in Labour, but episodes amplified through party factionalism and repeated by British media tainted progressives as anti-Semites. Conservatives within the party seemingly mishandled complaints of anti-Semitism specifically to weaponize them against leadership. Corbyn was eventually pushed out.

For his criticism of Israel's violence against Palestinians, Bernie Sanders, who lost family in the Shoah, faced frequent accusations of anti-Semitism from the right. When polls started showing Sanders's support was high among women, Latinx, and young Black communities, media cast his supporters as toxic white males. Finally, in the last weeks of the primary, fellow primary candidate Elizabeth Warren accused Sanders of saying a woman couldn't win the presidency. Media accepted the absurd charge as fact, Warren refused to correct the record, and liberals reveled in gullibility. On both sides of the Atlantic, campaigns proposing more equitable worlds were caricatured as bigoted.

None of these accusations was specifically or only a reaction against climate proposals. Instead, they occurred within a constellation of moderate intransigence. A serious climate response goes hand in hand with egalitarian social policies that liberals oppose. Party operatives depended on vague accusations tapping moral weakness or narcissism. The judgmental superego, Lacan said, takes a "senseless, blind character of pure imperativeness and simple tyranny."[17] In comments on the Corbyn smear campaign,

---

16  Richard Seymour, "Labour's Antisemitism Affair," *Jacobin*, April 6, 2018, jacobinmag.com.

17  Jacques Lacan, *Freud's Papers on Technique: The Seminar of Jacques Lacan, Book I*, ed. Jacques-Alain Miller, trans. John Forrester (New York: W.W. Norton & Company, 1991), 102.

Richard Seymour drew on this quote and painted a picture of the discursive strategy:

> You're accused of something that is grotesque, false and unjust. But, like the best smears, it contains just enough reference to real events that you have to answer it. If you defend yourself, you indict yourself, because then you're "denying the problem," "not taking it seriously," or "dismissing real concerns." If you don't defend yourself, you've admitted your guilt. If you attempt "nuance," it's evidence of equivocation or deviousness, and drowned out in the foghorn of nonsense. If you apologise, it's never good enough, it's always a fake non-apology, and at any rate confirms the original indictment. If you don't apologise, you just prove your guilt. If you take action, it's either not enough, or suspect, or proof that you were guilty in the first place. If you don't take action, you're a monster.[18]

That's the Freudian kettle logic of a false identity politics misused to support hierarchies and environmental exploitation, a sort of inverted mockery of the term's value in the Combahee River Collective statement. On both sides of the Atlantic, voters who'd put climate at the top of their concerns, fell for obvious slander. It's difficult to read the situation as anything other than enjoyment in deception. This, too, demonstrates Freud's observation about how, when the repressed surfaces, denial works by reality negation or guilt negation. If one is about to vote against the only decent option on climate change or a more equitable future, it helps to imagine one's opponents as vile.

Both in *Contribution to the Critique of Political Economy* and *The German Ideology*, Marx famously described a base-superstructure model for society. If the base of society is the (capitalist) material relations of production, what rises up on top of that base is an ideological, legal, political superstructure. Louis Althusser noticed the spatial metaphor implied by a superstructure rising on a base means both (1) whatever happens in the "upper floors" has at least some relative autonomy from the base, even though it grows out of the base, and (2) there is reciprocal action (superstructure acts upon the base and vice versa). That's why Althusser introduced ideological state apparatuses into this base-superstructure model

---

18 Richard Seymour, "Paralysis," Patreon, November 19, 2020, patreon.com.

as a way the state extracts compliance from people on the lower floors. Liberals love ISAs, especially if they shore up moral superiority. Liberals purify the symbolism of the superstructure without tampering with the base, but materialists know what Marx knew, that is, the superstructure corresponds to the mode of production at the base. If workers or renters are exploited by owners in day-to-day material relations of production, for example, it's no shock the legal and political superstructure favors owners. Anything else would be surprising!

Likewise, a legal and political superstructure build atop the particular exploitations unique to fossil fuels will react violently in concert when its base is threatened. What we need is a lever of power or a hinge point. We haven't yet found any.

## Gaia, or What Is to Be Done When Everything Will Be Fine?

In 1966, the scientist James Lovelock reported on global warming for Shell. He invented an electron capture detector for molecules at only a few parts per billion in the atmosphere, and while splitting his time between the private sector and Pasadena's Jet Propulsion Laboratory, he wondered whether the device might detect life from air samples on Mars. This research marked the start of his Gaia hypothesis.[19]

A couple of years later, he wrote, "If the atmosphere of the Earth is a biological contrivance, then it is reasonable to consider that the components are maintained at an optimum or near optimum composition."[20] Instead of Earth as a "space ship" or "life support system," Lovelock insisted on an older Greek deity: Gaia.

Over the years, he developed his Gaia hypothesis, especially in a 1975 article coauthored with a Shell manager named Sydney Epton. "Life exists only because the material conditions on Earth happen to be just right," but it wasn't a one-way street. "Life defines the material conditions needed for its survival and makes sure that they

---

19  For research on Lovelock's Gaia hypothesis, I am indebted to research by Leah Aronowsky, "Gas Guzzling Gaia, or: A Prehistory of Climate Change Denialism," *Critical Inquiry* 47, no. 2 (January 2, 2021): 306–27.

20  Ibid., 312.

stay there."[21] Gaia couldn't be wounded permanently. She nurtured her children, and her children returned the favor. In his 1979 book *Gaia*, he suggested our pollution was as natural as our breath. Perhaps Lovelock's plea for pollution was motivated reasoning. At the time, he worked for CFC supplier DuPont as an expert witness testifying to the chemical's safety. By the late eighties, a conference organized by the climatologist Stephen Schneider discussed the Gaia hypothesis, proving how far the concept reached.

In 1990, the television channel TBS aired a new superhero cartoon called *Captain Planet*. Each episode opened with narration: "Our world is in peril. Gaia, the spirit of the Earth, can no longer stand the terrible destruction plaguing our planet." Gaia sent five magic rings to five young "planeteers" representing continents and elements: Africa (earth), North America (fire), the Soviet Union (wind), Asia (water), and South America (heart). They harnessed the elements to fight polluting villains who are caricatures of vices (Hoggish Greedly, Verminous Skumm, Looten Plunder). When fights got too intense, Planeteers summoned eco-superhero Captain Planet. After crushing the week's polluter, Captain Planet exited with a message for viewers: "The power is yours!"

As Gaia turned into a pedagogical tool for children, she was adopted by the capitalist denial ISA. In 1995, Mobil ran an advertorial in the *New York Times* reassuring readers she'd take care of herself. "Good news: The end of the Earth as we know it is not imminent . . . to those who think industry and nature cannot coexist, we say show a little respect for Mother Nature."[22] Everything will be fine.

Today, we hear versions of "the kids will save us." Whatever the excesses of those who came before, the kids are all right, or the kids know what's up, or this generation isn't playing around, and so on. Generational politics never works out—class politics and material interests are the viable lens by which we analyze contradictions—but it's a comforting fantasy. If *Captain Planet* was aired to persuade young millennials to be environmentalists, today's youth climate movement certainly demonstrates a step beyond passive television consumption.

The youth climate movement benefits from "the kids will save

---

21 James Lovelock and Sydney Epton, as quoted in ibid., 316.

22 As quoted in Supran and Oreskes, "Assessing ExxonMobil's Climate Change Communications," 10.

us" wishes. It isn't hard to see how this illusion fits within broad liberal commitments. If generational politics works, then we needn't concern ourselves with class politics. Education and passion will get the job done without mucking up free markets with regulation or central planning, so the fantasy goes.

Greta Thunberg herself is rightly skeptical of performative concern from policymakers she castigates. Most famously, at the UN Climate Action Summit in September 2019, Thunberg, then sixteen years old, began her speech: "My message is that we'll be watching you." The audience cheered and clapped with the reflexive excitement of a bunch of pigeons spotting a morsel dropped on the ground. "This is all wrong," she continued. "I shouldn't be up here. I should be back in school, on the other side of the ocean. Yet you all come to us young people for hope. How dare you!"

Much as flagellants of the Black Death sought penance by whipping themselves bloody in displays of guilt, global capitalist climate governance enjoys moral masochism. And I want to be clear on this point, because I enthusiastically support the youth climate movement. What's characteristic of human guilt is that we'll apologize quickly to neutralize anxiety, even for things we haven't done, but we studiously avoid talk of whatever we secretly fear we've done. We hide from ourselves. When a young truth-speaker is invited to a UN summit, it's a safe bet audience members believe they've done nothing wrong. Her indictment, they imagine, is not directed toward them but toward another person at the conference, maybe in the next seat, but not themselves.

Or perhaps bourgeois liberalism appreciates the youth climate movement's nonviolence thus far. Andreas Malm rightly put the question as such: "At what point do we escalate?"[23] He pointed to John Lanchester's question about why direct action so far remained off the table:

> It is strange and striking that climate change activists have not committed any acts of terrorism. After all, terrorism is for the individual by far the modern world's most effective form of political action, and climate change is an issue about which people feel just as strongly as about, say, animal rights. This is especially noticeable when you bear in mind the ease of things like blowing up petrol

---

23   See Andreas Malm, *How to Blow Up a Pipeline* (London: Verso, 2021).

stations, or vandalising SUVs . . . So why don't these things happen? Is it because the people who feel strongly about climate change are simply too nice, too educated, to do anything of the sort? (But terrorists are often highly educated.) Or is it that even the people who feel most strongly about climate change on some level can't quite bring themselves to believe in it?[24]

Social progress in recent centuries was always accompanied by violence, at least so-called "violence" against property. Abolition of slavery, women's suffrage, defeating fascism, civil rights, the end of apartheid—all involved destruction. Frantz Fanon argued decolonization is always a violent phenomenon, and he spoke of a moral duty to silence the colonizer, "breaking his spiral of violence, in a word ejecting him outright from the picture."[25] Environmentalists often point to Erica Chenoweth and Maria J. Stephan's research claiming nonviolent resistance is the most successful strategy, but their book draws on a data set imposing "a criterion of one thousand annual battle deaths for inclusion in the intrastate conflict data set."[26] In this definition, a campaign that, say, kidnapped and killed 999 fossil fuel board members per year or firebombed their mansions with Molotov cocktails would not meet the criteria for violent intrastate conflict.

What are we to make, Malm asked, of 2050 net-zero pledges while new pipelines are still built with expected lifetimes of forty or sixty years? Committed emissions from currently operational fossil fuel infrastructure allowed to run to the end of its lifetime would inject 660 $GtCO_2$. Planned infrastructure would add a couple hundred more gigatonnes.[27] If you recall from earlier, the remaining carbon budget by 2025 should be at most only 193 $GtCO_2$.

---

24   John Lanchester, "Warmer, Warmer," *London Review of Books*, March 22, 2007, lrb.co.uk.

25   Frantz Fanon, *The Wretched of the Earth*, trans. Richard Philcox (New York: Grove Press, 2021), 35–6, 44.

26   Erica Chenoweth and Maria J. Stephan, *Why Civil Resistance Works* (New York: Columbia University Press, 2013), 258.

27   Existing infrastructure 660 $GtCO_2$ (460–890) and existing plus planned infrastructure 850 (600–1,100) per Intergovernmental Panel on Climate Change, *Sixth Assessment Report*, Working Group III, 16. See also Dan Tong et al., "Committed Emissions from Existing Energy Infrastructure Jeopardize 1.5°C Climate Target," *Nature* 572, no. 7769 (August 2019): 373–7.

"In a sense we all already know What Is To Be Done," said Gerry Canavan in an allusion to Lenin's pamphlet of the same name, alongside the question of whether violent resistance ought to have a place in ending emissions. "The real problem is navigating the grief and fear and selfishness that prevents us from ever actually putting what we all know to be true into practice."[28] If, as Walter Benjamin said, capitalism is a cult—"a pure religious cult, perhaps the most extreme there ever was," and "the first case of a blaming, rather than a repenting cult"—then it would be convenient if the solution were what liberals always holler: simply stop denying, and "believe the science."[29] Let the market solve its problems—maybe later when today's kids are grown-ups! That's bourgeois liberalism's hope, but where's the evidence for this strategy?

Liberalism disavows the hierarchies on which it perches. The liberal rejects the conservative's conviction that white men should rule over the rest, but its vision is not utopian or even anti-dystopian. The liberal feels the tug of egalitarianism but ultimately concludes all that socialist nonsense cannot work. Owners must rule over workers, and what is needed are market-based absolutions of a purity industrial complex for beautiful souls who care, who really, really care.

Drawing on Wilhelm Reich's study of fascism, in *Anti-Oedipus*, Gilles Deleuze and Félix Guattari said the bewildering question of political philosophy was always "Why do men fight *for* their servitude as stubbornly as though it were their salvation?" Reich suggested fascism came from sexual repression and authoritarian traits circulating within mechanized societies susceptible to violent fantasies. Withholding bread leads to the revolt of the hungry, Reich said, but the repression of sexuality produces a population that thinks they deserve neither sexual satisfaction nor even nourishment.[30] This was his formula for conservatism, a piece of

28   Gerry Canavan, "Of Course They Would: On Kim Stanley Robinson's 'The Ministry for the Future,'" *Los Angeles Review of* Books, October 27, 2020, lareviewofbooks.org.

29   Walter Benjamin, "Capitalism as Religion," in *The Frankfurt School on Religion: Key Writings by the Major Thinkers*, ed. Eduardo Mendieta, trans. Chad Kautzer (New York: Routledge, 2005), 259.

30   "The explanation is: The suppression of one's primitive material needs compasses a different result than the suppression of one's sexual needs. The former incites to rebellion, whereas the latter—inasmuch as it causes sexual needs to be repressed, withdraws them from consciousness

the puzzle, if not a fully satisfying answer. For Reich, mystical thinking is always reactionary, whether we speak of fundamentalist Christianity or astrology, of the appeal to tradition or faith in progress, of how Gaia will take care of herself or how the market will sort this out.

Living in the Global North entails poisoning the air with carbon dioxide, some of which will persist for hundreds of millennia and harm species that haven't yet evolved, simply by driving to get groceries. Fake solutions resolve guilt and negate moral culpability. It's a gift to bourgeois parties who know how to tamper with feelings, who know, as Marx said, how to position themselves as pragmatists speaking in the name of "property, family, religion, order." At every request for the slightest improvement, bourgeois parties cry foul. "Every demand of the simplest bourgeois financial reform, of the most ordinary liberalism, of the most formal republicanism, of the most shallow democracy, is simultaneously castigated as an 'attempt on society' and stigmatized as 'Socialism.'"[31]

They have no plan. Gaia cannot care for herself. Nothing is fine.

and anchors itself as a moral defense—prevents rebellion against *both* forms of suppression. In the consciousness of the average nonpolitical man there is not even a trace of it. The result is conservatism. Fear of freedom, in a word, reactionary thinking," Wilhelm Reich, *The Mass Psychology of Fascism*, ed. Mary Higgins and Chester M. Raphael (New York: Noonday Press, 1993), 31.

31 Karl Marx, *The Eighteenth Brumaire of Louis Bonaparte* (New York: International Publishers, 1990), 25.

# 10

# Adaptation and Mitigation, or Technological Seductions

*Like every generation that preceded us, we have been endowed with a weak Messianic power, a power to which the past has a claim. That claim cannot be settled cheaply. Historical materialists are aware of that.*

—Walter Benjamin

*Who will build the ark?*

—Mike Davis

2012: ExxonMobil CEO Rex Tillerson dismisses climate change as a workable problem for engineers. "Changes to weather patterns that move crop production areas around—we'll adapt to that. It's an engineering problem and it has engineering solutions," Tillerson says. "The fear factor that people want to throw out there to say 'We just have to stop this,' I do not accept."[1]

2015: Ahead of the Paris Agreement, the UN Global Compact announces a campaign to recruit one hundred companies to reduce emissions.[2] Pitched as "science-based targets," the press release highlights well-known brands like L'Oréal, H&M, and Nestlé. Now over a thousand companies pledge decarbonization by midcentury.

2018: Shell publishes the *Sky*, a public relations stunt showing "a technically possible, but challenging pathway for society to achieve the goals of the Paris Agreement." Never mind the IPCC

---

1   Matt Daily, "Exxon CEO Calls Climate Change Engineering Problem," *Reuters*, June 27, 2012, reuters.com.

2   "Campaign Launches to Close the Gap Between Corporate GHG Reduction Goals and a 2°C Scenario," Science Based Targets, May 20, 2015, sciencebasedtargets.org.

already games out such scenarios. Shell calls for a "change in consumer mindset" and upping carbon capture facilities from fifty to ten thousand. The document reads like a child wondering whether one of their three wishes from a genie could be unlimited wishes. Buried on the last page is a disclaimer: the company has no plans to move to a net-zero portfolio.

2021: Various outlets report General Motors, the largest auto manufacturer in the United States, will soon sell only electric vehicles. The *New York Times* reports GM will "phase out petroleum-powered cars and trucks," and *NBC News* says that "the automaker will go completely carbon neutral at all facilities worldwide by 2035." What GM actually says is it "aspires to eliminate tailpipe emissions from new light-duty vehicles by 2035." Significant caveats. Additionally, GM pledges carbon-neutral products and operations by 2040, a tall order when one doesn't control the electricity grid. It will add a few thousand fast charge stations, perhaps the company's effort to catch up to Tesla, which has ten times as many. Even if the company reaches all its goals, its vehicles will emit long past the 2035 date for which it reaps praise now.

## Adaptation

John Maynard Keynes rightly said, "Anything we can actually do we can afford." But without a sober assessment of staggering costs and physical limits to our mitigation options, as a denial device the technological solutions seduce us into believing capitalism will finally pay its bill. Just as Lacan warned that "truth is firstly a seduction, intended to deceive you," what might save us later is what obstructs us now.[3]

There are two broad responses to climate change: adaptation and mitigation. Mitigation comprises a few broad categories (efficiency, electrification, carbon capture), while adaptation encompasses an expansive menu of strategies to cope with what's about to happen. This chapter examines options with a watchful eye to how necessary measures double as denial in the near term by hinting, more or less, that everything will be fine. A third option is suffering.

---

3 Jacques Lacan, *The Other Side of Psychoanalysis: The Seminar of Jacques Lacan, Book XVII*, ed. Jacques-Alain Miller, trans. Grigg Russell (New York: W. W Norton & Company, 2007), 185.

Adaptation is adjustment to new normals: train firefighters, weatherize infrastructure, change agricultural and forestry management, prepare for infectious diseases, allocate funds for flood insurance, plan for destruction of rails and road, and even for the relocation of tourist destinations. Yes, one IPCC report suggested "shifting ski slopes to higher altitudes and glaciers; artificial snow-making."[4] The most recent report cautioned against maladaptation, such as when fire or flood prevention measures disrupt ecosystems and make matters worse.

We will build seawalls. That much is certain (at least where there's money). Whether we'll build them in time or after weakly managed retreats of shoreline property is another matter. Beaches are only a couple of meters high at the base. Depending on the SSP scenario, sea level will rise anywhere from half a meter by end of century to sixteen meters by 2300. Sea level rises two to six meters over the next two millennia if we limit warming to 2°C. Add another degree and it jumps to between four and ten. At 5°C, it could be as much as twenty-two meters.[5] Globally, four in ten people live within a hundred kilometers of a coast.

One estimate tallied up $400 billion needed for seawalls for coastal communities in the United States. Florida topped the vulnerable states. The price per affected person in many communities will be more than $100,000.[6] Seawalls are planned along the Potomac River in Washington, DC, and New York City's real estate is judged so valuable it's worth a six-mile seawall estimated to cost $119 billion.[7] Imagine how the police will panic over the tamest protest if a discreet bomb lodged at a weak point could drown lower Manhattan.

Indonesia is moving its capital from Jakarta to Borneo. What of the inhabitants of Jakarta, which could be the world's most populous city before long? One group absurdly proposed damming the English Channel and building a vast wall from Scotland to

4   Intergovernmental Panel on Climate Change, *Climate Change 2007 Synthesis Report*, 2007, 57.

5   Intergovernmental Panel on Climate Change, *Sixth Assessment Report*, Working Group I, 2021, 22, 79.

6   "Study: U.S. Coastal Communities Face More than $400 Billion in Seawall Costs by 2040," Institute for Governance & Sustainable Development, June 20, 2019, igsd.org.

7   Anne Barnard, "The $119 Billion Sea Wall That Could Defend New York . . . or Not," *New York Times*, January 17, 2020, nytimes.com.

Norway.[8] Mark Fisher repeated the aforementioned dictum that "it's easier to imagine the end of the world than the end of capitalism," but now it's easier to imagine abandoning capitals or taming the North Sea than the end of capitalism.[9]

Seawalls take time to build. It's not certain we'll feel sufficient urgency. In South Korea, the world's largest seawall took a couple of decades to build, while public outcry and legal challenges interrupted construction over the way the wall affected homes, livelihoods, animal habitats, and ocean views. Will property owners, especially the useless vultures kindly called landlords, along a metropolitan harbor or a scenic beach town permit construction if it ruins the aesthetics and depreciates prices now when sea level rise is still decades or centuries off?

More likely, as the waters gradually rise, districts will flood here and there, eventually more frequently. The community will ask how officials let this happen. A Great Dithering.

At COP26 in Glasgow, Barbados prime minister Mia Mottley criticized wealthy nations for falling short on adaptation funds pledged to developing nations. "Failure to provide the critical finance, and that of loss and damage, is measured, my friends, in lives and livelihoods in our communities. This is immoral, and it is unjust," she said to closed ears. She pointed to Covid-19 vaccine hoarding and asked, "Can there be peace and prosperity if one-third of the world literally prospers and the other two-thirds of the world live under siege and face calamitous threats to our well-being?" In her closing litany of states that would be drowned, burned, and starved, she took a prophetic turn from the Book of Isaiah:

> For those who have eyes to see, for those who have ears to listen, for those who have a heart to feel, one point five is what we need to survive. Two degrees is a death sentence for the people of Antigua and Barbuda, for the people of the Maldives, for the people of Dominica and Fiji, for the people of Kenya and Mozambique, and, yes, for the people of Samoa and Barbados. We do not want that

8  "Protecting Europe with Dams across the North Sea," MS Amlin, July 7, 2020, msamlin.com; and Claire Moses, "As Sea Levels Rise, Scientists Offer a Bold Idea," *New York Times*, February 14, 2020, nytimes.com.

9  See Mark Fisher, *Capitalist Realism: Is There No Alternative?* (Winchester, UK: Zero Books, 2009).

dreaded death sentence, and we've come here today to say, "Try harder! Try harder!"

During that meeting, the foreign minister of the Polynesian island nation of Tuvalu spoke while standing knee-deep in the ocean. Tuvalu's highest point is four and a half meters above sea level. Eleven thousand live on the island. Do their legal rights and maritime zones evaporate if other states raise temperatures? Can the people of Tuvalu relocate and maintain sovereignty once the land is submerged?

I write in a café across the worn road from Baltimore Harbor. The port was established before colonists declared independence, and the cobblestone and brick streets in Fell's Point have been here since the eighteenth century, when the area was built up for shipping. Last week, the water was much higher, on account of rainfall. Sometimes, the area floods in severe storms. It will start happening more often, slowly at first, until the brackish water spills over. "It didn't use to be like this," people will reminisce as the ancient streets once strolled by Frederick Douglass and Edgar Allan Poe go under more frequently each year.

## Mitigation and Decarbonization Pathways

Mitigation is the prevention or reduction of emissions. Since some carbon dioxide will stay in the atmosphere for many thousands of years, a portion of emissions are effectively permanent. Aside from limited uses for biomass and hydrogen, there are three broad decarbonization pathways: efficiency, electrification, and carbon dioxide removal.

First, optimizing processes for efficiency is the simplest pathway. We fill our landfills with single-use plastic. We construct buildings with far more concrete than is necessary. We toss food. We run dishwashers and lights in the primetime hours when we cannot generate electricity from the sun. We use individual vehicles instead of public transport. Efficiency or reduction is straightforward in theory. Roadblocks tend to be social. Arguably, the biggest eco-conscious behavioral change so far was recycling, and it's still so confusing and inefficient that only 8 percent of plastics are recycled (roughly twice as much is burned, releasing stored carbon as carbon dioxide).

Second, electrification removes needs for fossil fuels. Electric vehicles and stoves are easy, but other processes will be complicated. If you recall our discussion of industry emissions in chapter 4, steel production usually involves coal-fired furnaces that heat iron oxide and split off oxygen by bonding it with carbon, creating two sources of carbon dioxide for one product. If we electrified the heat process and bonded the oxygen with hydrogen, and if we boosted recycling efficiency, we could hold steady at our current amount of total steel even as developing nations demand more. At the same time, we could cut carbon dioxide from the steel industry. Any emissions remaining might be subject to the third decarbonization pathway.

Third, carbon capture or carbon dioxide removal (CDR) refers to several negative emissions technologies. If we strip carbon from the air, either by scrubbing flue gasses at factories and power plants or with machines pulling directly from the air, then we can continue emitting in hard-to-decarbonize industries without permanent damage. These technologies are in their infancy and currently expensive. Heavy use of carbon capture and storage is the only conceivable pathway to a 1.5°C future, though we'd still overshoot that threshold before coming back down with sufficient carbon capture.

In 2021, the Global CCS Institute reported sixty carbon capture facilities operational in the world with a combined sequestration capacity of 40 $MtCO_2$ per year.[10] Some of what is captured goes into enhanced oil recovery, the fertilizer industry, alternative fuels, and other processes rather than storage. More than a hundred projects were in some stage of production. The vast majority of that capacity is integrated with chemical or power plants. Just shy of 0.01 $MtCO_2$ per year is sequestered by nineteen direct air capture facilities worldwide.[11]

Affixing mechanisms to collect flue gasses at factories is how we will likely decarbonize a certain amount of the industrial sector. With cement kilns, this can be done simply via carbon dioxide capture at the exhaust stack. Capturing gasses will be more complicated for traditional steel due to the sheer size of steel mills and the many locations from which gasses are emitted.

---

10  "Facilities," Global CCS Institute, accessed May 28, 2022, co2re .co; and "Global Status of CCS 2021," Global CCS Institute, 2021.

11  "Direct Air Capture," International Energy Agency, 2021, iea.org.

Shell and ExxonMobil promote carbon capture because the fantasy of sin without consequences is critical for their business model. To them, CDR denies the need to keep fuels in the ground. In the coal industry, one of several meanings of "clean coal" (an oxymoron) is a power plant built "carbon capture–ready" where owners advertise it as able to capture carbon were the equipment to be installed. Recall from chapter 2 that Delta advertised carbon capture not as something it did but instead what other companies might do someday.

When climatologists tell us carbon capture is the only pathway, in conjunction with steep emission declines, to meet certain targets, they do so with full knowledge of how unlikely these possibilities are under current political economies. Their job is to research options. The denialist ISA promotes carbon capture for different reasons, and the cost is always omitted. Remember Tillerson panned climate change as "an engineering problem." You can detect more than a hint of this arrogance in capitalist climate governance as well. In effect the technologies on which we will depend are a truth spoken with the intent to deceive. Until emissions drop to 10 to 15 percent of current levels, CDR will not suffice. Let's discuss options for carbon dioxide removal.

## CDR: Trees

Can we not simply plant trees? We've already seen the limits of afforestation as a carbon offset. Say a tree absorbs 22 $kgCO_2$ per year. It would take more than six hundred trees to absorb Americans' annual emissions of 14.5 $tCO_2$, and there are physical limits to how many we can plant. There are currently four billion hectares of forest in the world. A recent study in *Science* suggested we could boost forests by an additional billion hectares and sequester 205 GtC.[12] That would drop carbon dioxide by 50 ppm and reverse damage by only a couple decades.

There's a trick to remember when calculating CDR strategies. Simply put, to eliminate a tonne of carbon dioxide from the atmosphere, we actually need to remove two tonnes.[13]

---

12    Jean-François Bastin et al., "The Global Tree Restoration Potential," *Science* 365, no. 6448 (July 5, 2019): 76–9.

13    Per Zeke Hausfather, Twitter, February 12, 2022, twitter.com.

Removing a tonne of carbon dioxide via any CDR strategy is effectively the same as not emitting that tonne in the first place. But when a tonne is emitted, it doesn't all remain in the air. After ocean and land uptake, 44 percent stays in the atmosphere. In higher future emissions scenarios, a greater portion stays in the atmosphere, so for simplicity's sake, modelers say roughly half stays in the air. When levels drop in the atmosphere, the ocean and land will outgas to reach equilibrium with the air. That's good news since ocean outgassing into the atmosphere (where we know how to scrub out emissions) means we can reverse ocean acidity by removing carbon dioxide from the air. In short, to reverse emissions where less than half stays in the atmosphere, any CDR technique must remove twice as much carbon dioxide as we want to eliminate from the atmosphere.

What does this calculation mean for that upper limit of 205 GtC removed by planting every tree possible? First, in a moderate emissions scenario like SSP2-4.5, carbon dioxide will hit 600 ppm by 2100.[14] Second, using the trick for calculating removal, where what the trees pull from the air reduces concentration only by half as much, trees would eliminate about 100 GtC. One ppm equals 2.13 GtC, so divide 100 GtC by 2.13 to get a number just shy of 50 ppm. It isn't much when concentration currently rises ~2.5 ppm per year. So planting every tree possible in the entire world might reverse fifteen or twenty years of damage.

This is far too little to hit our goals. In moderate scenario SSP2-4.5, this strategy alone would still leave us with an atmosphere at 550 ppm, roughly twice the pre-industrial average. Or in unabated scenario SSP5-8.5, where concentration hits 1,135 ppm by end of century, removing around 50 ppm means even less.[15]

Trees are the cheapest CDR technique. Authors of the study above figured the cost to cover an extra 11 percent of Earth's land with a trillion new trees is only $300 billion.[16] Afforestation removes carbon dioxide at a cost of fifteen dollars per tonne, by far the cheapest option.[17] However, monetary costs aren't the only costs.

---

14   Malte Meinshausen et al., "The Shared Socio-Economic Pathway (SSP) Greenhouse Gas Concentrations and Their Extensions to 2500," *European Geosciences Union* 13, no. 8 (August 13, 2020): 3571–605.

15   Ibid., 3589.

16   Damian Carrington, "Tree Planting 'Has Mind-Blowing Potential' to Tackle Climate Crisis," *The Guardian*, July 4, 2019, theguardian.com.

17   "Fifteen dollars" per Zeke Hausfather, Twitter message to author, February 18, 2022.

Afforestation costs in soil nutrients as well as water. In a study of CDR techniques, the researchers Pete Smith et al. warned trees could dampen Earth's albedo effect. Crops and grasses reflect more than forests. In northern latitudes, trees planted on land that's currently flat and snow-covered means branches poking through, making the region darker. Afforestation in northern latitudes yields a neutral or even net warming effect.[18]

Maximum afforestation could use up to a fifth of total agricultural land.[19] More land for trees means less land for cities, railways, solar panels, and crops. The journalist David Wallace-Wells noted, "Planting forests at a scale large enough to meaningfully alter the planet's carbon trajectory, for instance, could elevate food prices by 80 percent."[20]

Finally, trees can only draw down carbon dioxide so long as they find water and other nutrients in adequate proportions. The Grantham Institute's Bonnie Waring likened the problem to proportions of ingredients when baking a cake. Trees gladly consume available carbon dioxide until there is no longer enough nitrogen or phosphorus to add to the mix.[21] After this threshold, trees cannot absorb additional carbon dioxide.

So we should plant more trees, of course. Trees enrich the world aesthetically and benefit animals. Afforestation will help with carbon uptake. But afforestation is not a panacea. It's a tool among many, the most important of which will still be eliminating emissions in the first place. To drop concentration 50 ppm, planting the absolute maximum number of trees would be insufficient; doing so would only reverse a couple of decades' emissions. When companies promise to plant a tree for every purchase, they are asking you to stick your head in the sand to soothe guilt. Of course, wildfires rip through corporate forest offsets each summer. How often do companies relying on these offsets notify customers that they weren't carbon neutral this year after all, even by their own tortured accounting?

---

18  Smith et al., "Biophysical and Economic Limits," 42–50.

19  Ibid., 5.

20  David Wallace-Wells, "After Alarmism," *New York Magazine*, January 19, 2021, nymag.com.

21  Bonnie Waring, "There Aren't Enough Trees in the World to Offset Society's Carbon Emissions—and There Never Will Be," The Conversation, April 23, 2021, theconversation.com.

## CDR: Bioenergy with Carbon Capture and Storage

Bioenergy with carbon capture and storage (BECCS) uses carbon capture devices we can affix to power plants, but instead of burning fossil fuels, it burns biomass. First, organic materials like wood pellets or crops are collected and, with them, carbon. Second, biomass is burned, and its energy converted to heat, which can be turned into electricity. Third, the resulting carbon dioxide is captured and stored underground.

Advantages of BECCS include production of renewable energy, and the feedstock draws carbon dioxide from the environment. Ideally, emissions are captured. Even if emissions are not captured, the carbon dioxide released into the air is roughly the same quantity that was pulled from the air during the feedstock's lifetime, keeping bioenergy carbon-neutral. With BECCS still in its infancy, there are only a few plants now.

BECCS won't affect albedo, and its monetary cost is already near where we hope direct air capture can be in several decades.[22] But like afforestation, feedstock biomass consumes soil nutrients and water. BECCS will require lots of fertile land, so the chief concern is cropland lost to biomass harvesting for a low EROI.

The United States and EU's use of biofuels like ethanol likely contributed to the high food prices that drove desperate families into urban centers at the onset of the Syrian civil war. As the war began, Oxfam reported 40 percent of US corn went to fuel, not food, and the amount of land the EU devoted to biofuels could have fed 127 million mouths.[23] BECCS had nothing to do with this, but we should be clear that global capitalist climate governance, as a bourgeois management project, might prioritize a low EROI renewable over the starved people of the world.

In 2021, Australia concocted an emissions-reduction pathway dependent on BECCS. Billed as "technology not taxes," the plan would remove 15 percent of greenhouse gasses by midcentury through BECCS. It was a farce, and the University of Melbourne researcher Kate Dooley noted it would eat fully 6 percent of all Australia's land for biomass harvest. The Grattan Institute director Tony Wood underscored the irony of a government promoting

---

22  Smith, "Biophysical and Economic Limits."

23  Sarah Kalloch, "Burning Down the House," Oxfam, August 4, 2012, politicsofpoverty.oxfamamerica.org.

a costly technology as a solution without investment. "The report says it's based on an assumption the technology is not economically viable in absence of incentives," Wood cautioned. "But there's nothing in the report that describes what incentives would trigger such an investment."[24] So long as BECCS remains a theoretical possibility, it can be used in projections without the political risk of actual investment and implementation.

Monetary cost estimates vary widely for BECCS, from fifteen to four hundred dollars per tonne of $CO_2$. It should go without saying that two options always threaten to dramatically slash the price: simply not capturing the carbon dioxide or using it for "enhanced oil recovery," as is the case for much of our existing carbon capture infrastructure. IPCC scenarios feature BECCS as an important pathway to decarbonization, but companies could sell the product as an extraction aid to oil companies.

## CDR: Direct Air Carbon Capture and Storage

Direct air carbon capture and storage (DACCS) removes carbon dioxide from the air using chemical absorption and mineralization. Ideally, the product is mineralized underground. Carbon dioxide capture isn't new. That ability makes possible travel on submarines and life in space. Remember that some power plants already capture emissions.

The Swiss company Climeworks is a global leader in direct air capture and is even sketching out potential ocean carbon removal. Its modular collectors draw in air through filters absorbing carbon dioxide. After the filter is saturated, the collector closes, and a heat process captures concentrated carbon dioxide. The end product can be used for fuels and materials, or it can be stored. Since Climeworks sells its services as offsets (for corporations, individual indulgences, or even as eco-friendly gifts), it may well be net neutral.

In fall 2021, Climeworks's Orca plant switched on in Iceland. It was the largest direct air capture plant in the world, instantly boosting global carbon capture capacity by half overnight. Over

24  Graham Readfearn, "'A Farce': Experts Dismiss Government Claims a Controversial and Unproven Technology Will Cut Emissions by 15%," *The Guardian*, November 18, 2021, theguardian.com.

the course of a year, Orca captures 4,000 $tCO_2$, the equivalent of 870 cars. "If it works," said the climate scientist Peter Kalmus, "in one year it will capture three seconds worth of humanity's $CO_2$ emissions . . . at incredible expense. I'm rooting for it, but only a fool would bet the planet on it." If we took no other mitigation measures, we'd need ten and a half million Orcas to zero out current emissions.

At the moment, Climeworks charges six hundred dollars per tonne of carbon dioxide removed. This boutique industry's machines are built by hand, so mass production will draw down the price point. Across the industry costs are dropping quickly, and we expect the price to hit two hundred dollars by 2030 and then one hundred by 2040 or 2050. Some in the industry hope for fifty dollars.

Bill Gates has also pointed to this off-in-the-distance hundred-dollar price. He funds a carbon capture company called Carbon Engineering. Its chief scientist, David Keith, worked out a range for direct air capture costs from $94 to $232, but the lower number is for an option where carbon dioxide is turned into a fuel, not stored. Costs are higher if carbon capture and storage actually includes storage in perpetuity.

This method faces few downsides apart from electricity and monetary costs, both of which are astronomical. If Orca scrubbed the emissions of an average US household, it would take more than double the electricity the home uses.[25] Nevertheless, it is the cost in money that is truly eye-popping.

Taking the aspirational low price of one hundred dollars per tonne, how much would this cost? At our current emissions of 41.4 $GtCO_2$ per year, the cost to clean up the mess is over $4 trillion per year. But to reach energy equilibrium, we can't just zero emissions. We must go further and drop concentrations to 350 ppm.

In the moderate scenario SSP2-4.5, where atmospheric concentration hits 600 ppm by 2100, cleaning up emissions with DACCS alone would cost $391 trillion or $5 trillion per year if paid out over the remaining three-quarters of a century. That's a trillion and

---

25  Climeworks estimates the electricity cost is 650 kilowatt hours (kWh) per tonne removed. An average US household of 2.5 persons emits 36 $tCO_2$ per year. (Note: this is for total emissions all persons in that household, not just electricity, so the comparison is not perfect.) Removing those emissions with Orca would take 23,000 kWh electricity to remove, whereas an average 30 kWh per day home consumes about 11,000 kWh of electricity.

a half dollars for every part per million. In the high emissions scenario, the bill comes to $1.23 quadrillion or $16 trillion per year.[26]

These are absurd numbers. The costs are so astronomically unthinkable many environmentalists' blood boils at the mention of carbon capture and storage. Carbon capture experts or climatologists who think DACCS has a limited role—I count myself among those who want to see billions or preferably trillions dumped into this technology—will tell you it's only one option among mitigation strategies. At the same time, earlier in the book I suggested one hundred dollars per tonne would be under a dollar per gallon of gasoline if passed onto the consumer as a surcharge.

Building on the philosopher Olúfémi Táíwò's case for climate reparations, journalist David Wallace-Wells argued direct air capture could be the instrument.[27] At a hundred dollars per tonne, tallying up historic emissions by nation, the bill is straightforward: $50 trillion owed by the United States, $30 trillion from China, $8 trillion from the UK, and in total $250 trillion.

It is difficult to imagine the lords paying for what they've done. It's easy to imagine legitimate and critical solutions in the future amounting to grifts and dithering today.

---

26 I'm running numbers as if all removal were done through carbon capture, though natural sinks would assist some until sinks weaken. In the moderate emissions scenario SSP2-4.5, where we hit 600 ppm by 2100, we'd need to remove 250 ppm over the rest of the century. Doing a few calculations, first, if 1 ppm equals 2.13 GtC then we can convert 250 ppm to 533 GtC. Second, carbon is 12/44ths carbon dioxide, so the weight of those 533 GtC is also 1,954 $GtCO_2$. Third, to lower atmospheric concentration of carbon dioxide by a target amount we need to remove double that amount, so we'd need to remove 3,909 $GtCO_2$ by 2100 (notice this is in the vicinity of emissions from burning all fossil fuels from charts in chapter 6). In this moderate emissions scenario, at one hundred dollars per tonne we'd need to pay $391 trillion. In the high emissions scenario SSP5-8.5, the concentration of 1,135 ppm by 2100 means we need to remove 785 ppm to get back to equilibrium at 350 ppm. Converting to carbon dioxide, then doubling gets us 12,261 $GtCO_2$, which would cost $1.23 quadrillion. See Malte Meinshausen et al., "The Shared Socio-Economic Pathway (SSP) Greenhouse Gas Concentrations and Their Extensions to 2500," *Geoscientific Model Development* 13, no. 8 (August 13, 2020): 3571–605; and Intergovernmental Panel on Climate Change, *Sixth Assessment Report*, Working Group I, 2021, 80.

27 David Wallace-Wells, "The Case for Climate Reparations," *Intelligencer*, November 1, 2021, nymag.com.

## Low-to-Medium Confidence

Many minor mitigation techniques exist as well. For a long time, interest grew around fertilizing the ocean with iron to stimulate phytoplankton and increase carbon dioxide uptake, but it now looks disappointing. Others suggest enhancing Earth's natural rock-weathering process by mixing carbon dioxide with minerals like olivine or calcium silicates, while another weathering option would apply olivine powder on forests or farmland to boost rain's natural weathering capacity. Or we could improve agriculture practices, since 13 percent of carbon dioxide and 44 percent of methane come from agriculture, forestry, and other land uses.[28]

More options include soil carbon sequestration (improves agricultural management), biochar (burns biomass under anoxic conditions and adds the resulting charcoal to soil), peatland restoration, artificial ocean upwelling (pumps nutrient-rich deep water to the surface to increase carbon uptake), restoration of coastal ecosystems, and ocean alkalinization (increases carbon dioxide uptake with deposition of alkaline minerals).

My skepticism is shared by the scientific community. In the IPCC's *Sixth Assessment Report*, all CDR options are tagged with low or medium confidence. There is no clearly superior option.[29]

## Solar Radiation Management

In *The Planet Remade*, Oliver Morton recounted a couple of questions reverberating through a hall at the University of Calgary, where attendees had convened to discuss geoengineering. The physicist Robert Socolow asked two questions: "Do you believe the risks of climate change merit serious action aimed at lessening them? Do you think that reducing an industrial economics carbon-dioxide emissions to near zero is very hard?"[30]

Reactionary climate deniers answer no/yes. Those selling easy solutions answer yes/no. Though he's far more bullish on solar radiation management than I am, Morton's answers match mine: yes/yes.

---

28  Intergovernmental Panel on Climate Change, *Sixth Assessment Report*, Working Group I, 2021, 188.

29  Ibid., 756–7; and ibid. (pre-print), 5SM-18–19.

30  Oliver Morton, *The Planet Remade* (Princeton: Princeton University Press, 2016), 1.

We are in an emergency. Geoengineering with CDR or a veil to reduce sunlight might buy time. Or in the wildest of technocratic fantasies, veil-making could tailor the planet to our preferences. Left environmentalists are rightly skeptical of options permitting unabated emissions. We may be forced to reconsider and weigh the known barbarism of warming against the unknown catastrophes of a tailored Earth.

The name "solar radiation management" caught on as a joke. After the chemist Paul Crutzen's 2006 paper in cautious defense of geoengineering, researchers took up the taboo topic again after years of dormancy. The Italian physicist Cesare Marchetti first published the word "geoengineering" in 1977. Around the same time, the Russian climatologist Mikhail Budyko suggested cooling the atmosphere by flying jets high up with sulfur-enriched fuel. After Mount Pinatubo erupted in 1991, a National Academies of Science report considered injecting sulphates into the atmosphere or launching mirrors into orbit. In 1997 the physicist and "father of the hydrogen bomb" Edward Teller, along with Lowell Wood and Roderick Hyde, published a paper on stratospheric veilmaking. But it was Crutzen, the man who won a Nobel Prize for discovering how pollutants could destroy the ozone, who broke the taboo.

The NASA Ames Research Center hosted a conference on geoengineering after Crutzen's paper stirred interest. But "geoengineering" was still too controversial to fly with the higher-ups. Another name was needed. In Morton's history, one of the organizers, the atmospheric scientist Ken Caldeira, "responded in partial jest that they should just replace 'geoengineering' with the dreariest management-speak circumlocution they could come up with—'something like "solar-radiation management."'"[31] The abbreviation SRM stuck.

The basic idea behind geoengineering via solar radiation management is simple. The sun continuously strikes Earth with 174,000 TW; 30 percent is reflected, and 123,000 TW strikes the surface at 240 $W/m^2$. Prior to industrialization, Earth radiated back out an equal proportion of energy but now absorbs excess energy. Whereas other mitigation strategies reduce absorption back to preindustrial rates with emissions cuts or carbon capture, SRM tackles the problem from the other direction by turning down the sunlight entering the system in the first place.

---

31   Ibid.,156.

If we dialed down inbound radiation, then we'd solve the energy imbalance problem without eliminating carbon dioxide emissions. Or more realistically, SRM, alongside other mitigation strategies, could limit harm and buy time while we decarbonize.

Planes or balloons would ascend into the stratosphere and release aerosol payloads. A 2010 Aurora Flight Sciences report commissioned by David Keith proposed a fleet of fourteen Boeing 474s. The Aurora report proposed building a few dozen specialized aircraft for $7 billion and operating at $2 billion annually.[32] Flying so high is dangerous. Safer options include floating balloons high in the stratosphere and pumping up aerosols. Obstacles for balloons include maintaining fluid pressure and keeping the pipes from twisting. Whatever the challenges, such a program would cost only a few billion dollars per year. SRM is not yet heavily incorporated in integrated assessment models, but once it is, it will lure policymakers with a cheap fix.

The canonical example for SRM's potential was Mount Pinatubo's eruption in 1991, which pumped enough aerosols into the stratosphere to drop temperatures 0.5°C for two years. Pinatubo also provided warnings. Cooling wasn't uniform. Neither were decreases in rainfall. In *This Changes Everything*, Naomi Klein pointed to a 20 percent reduction in rain across southern Africa and a 10–15 percent reduction in South Asia. The drought hit as many as 120 million people and crop losses ranged from 50 to 90 percent. In the moment, Klein recalled, "few linked these disastrous events to the Pinatubo eruption since isolating such climate signals takes time."[33] Would the Global North notice or mind if SRM starved so many in the Global South? The lesson of Syria's drought and its relation to global warming and US and EU ethanol suggests we in the Global North might transform the environment carelessly.

By blocking some radiation driving the hydrological cycle, an Engineered Planet is drier than a Baseline or Greenhouse Planet. If you think of our world today as a Baseline Planet, a warmer world as a Greenhouse Planet, and an SRM-veiled world as an Engineered Planet, you can compare costs and benefits. On our Baseline Planet, sulfur dioxide plays the biggest role in negative forcing that masks warming. If you recall from chapter 5, aerosols

---

32  Ibid., 102–6.
33  Klein, *This Changes Everything*, 272.

like sulfur dioxide were the reason Stephen Schneider and S. Ichtiaque Rasool suggested global cooling was possible back in 1971,
before carbon dioxide concentration was understood to be far more
determinative. Sulfur dioxide masks 0.5°C of the warming we'd
experience otherwise.[34] Aerosols in the troposphere lead to lung
problems and acid rain, but if we cleared up pollution the planet
would get hotter. An SRM Engineered Planet offers a solution. In
our Baseline Planet's troposphere where weather occurs, sulfur
dioxide washes out quickly with rain. However, in the stratosphere a smaller quantity of aerosols achieves the same cooling
effect and remains for a couple of years. "So if you put just 4 per
cent of today's total sulphur emissions into the stratosphere rather
than the troposphere," Morton calculated, "you could get rid of
all the other 96 per cent, along with all the damage that they do
to human health, while preserving the cooling they had until then
provided."[35] We'd eventually inhale falling aerosols but much less
than we inhale today.

One obvious criticism of SRM is that if it makes continued
emissions tolerable, carbon dioxide still acidifies the oceans of the
Engineered Planet. More acidic oceans along with rising waters
means a toxic environment and less sunlight for the quarter of
marine species depending on shallow coral reefs. Reefs will mostly
die around 1.5–2°C but may also die if we limit warming but not
acidity. This is not hypothetical. In addition to more than 90 percent
of the added heat going into the ocean, three-eighths of carbon
dioxide emissions end up there. Acidity already damages carbonate
shells of small organisms like phytoplankton, killing off the base
of the food chain.

Larger organisms can survive acidity but not starvation. In 2016,
a four-year-long study on acidification found that a phytoplankton
named *E. huxleyi* slowed down its calcifying process under highly
acidic conditions, meaning it produced fewer scales on its body.[36]
In 2021, another study found the shells of microscopic pteropods
were 37 percent thinner in acidic water.[37] The IPCC says SRM

34  Intergovernmental Panel on Climate Change, *Sixth Assessment
Report*, Working Group I, 2021, 7.

35  Morton, *The Planet Remade*, 280–1.

36  Sean Greene, "The Damage Wrought by Acidic Oceans Hurts More
than Marine Life and Lasts Longer than You Think," *Los Angeles Times*,
July 8, 2016, latimes.com.

37  NOAA, "Acidification Impedes Shell Development of Plankton

attempts to mask a problem rather than "attempting to address the root cause of the problem, which is the increase in GHGs in the atmosphere . . . SRM does not address other issues related to atmospheric $CO_2$ increase, such as ocean acidification."[38] CDR could address this problem by drawing carbon dioxide out of the air such that the oceans outgas into the air where we can capture it. However, we're now back to the sheer cost of carbon capture. If the chief benefit of SRM is a cheap price that still requires outrageously expensive carbon capture, what's the point?

An Engineered Planet could be tailored to the same average temperature as a pre-industrial Baseline Planet, but average isn't everywhere. Temperatures would be distributed differently. The poles would be warmer and the tropics cooler on an Engineered Planet. If we wanted to protect the ice caps by restoring polar temperature, the tropics would need to be set much cooler. In reality, we might work around this problem by injecting lots of aerosols at the poles and less at the tropics. This method, as would any pattern of injection, will wreak havoc on the hydrological cycle.[39] Simulations indicate that while the Engineered Planet is drier, precipitation decreases over the oceans and increases over the northern hemisphere, and areas that are now wet would be less so. There are deeply serious concerns about our potential to trigger droughts and eliminate monsoons. Models suggest SRM will protect high elevations in Africa from malaria, a pathogen sensitive to temperature, but increase transmission in sub-Saharan Africa and southern Asia.[40] Implementation will hit snags not forecasted in models. As the weather forecaster David Titley once said, "All models are wrong; but precipitation models are particularly wrong."[41]

"With geoengineering there is no magic bullet. You will improve conditions in one part of the planet, but they will deteriorate in others," said the climate modeler Jim Haywood. "There are winners and losers. Then you get into the whole moral and ethical

Off the U.S. West Coast," NOAA Research News, accessed May 4, 2022, research.noaa.gov.

38  Intergovernmental Panel on Climate Change, *Sixth Assessment Report*, Working Group I, 2021, 105.

39  Morton, *The Planet Remade*, 114–5.

40  Colin Carlson et al., "Solar Geoengineering Could Redistribute Malaria Risk in Developing Countries," *Nature Communications* 13, no. 2150 (April 20, 2022).

41  David Titley as quoted in Morton, *The Planet Remade*, 116.

deliberation—who chooses?"[42] We could theoretically increase rainfall in the Sahel, but how likely is that option if it boosts hurricanes pushing through the Gulf of Mexico? Who chooses, indeed. Bill Gates dumps millions into solar radiation management research such as David Keith's Stratospheric Controlled Perturbation Experiment (SCoPEx). Gates called SRM a "Break Glass in Case of Emergency' kind of tool" that would "raise thorny ethical issues."

The very notion of tailoring an Engineered Planet, when the tailors will likely be persons or state apparatuses with weapons and capital historically deployed to racist effect, fills me with pessimism. There's no reason why we must inject aerosols only in one hemisphere or another, but excuses aren't difficult to imagine. Klein suggested that "it is all too easy to imagine scenarios wherein geoengineering could be used in a desperate bid to, say, save corn crops in South Dakota even if it very likely meant sacrificing rainfall in South Sudan."[43] Suppose we learn to manipulate certain areas to our preference but must treat other regions as sacrifice zones. Under what circumstances would millions throughout the Sahel or the Indian subcontinent, threatened with drought or a missing monsoon season, outweigh the interests of farmers in Iowa?

Once under way, SRM would require permanent implementation until carbon dioxide concentration dropped. If, for any reason, we suspended an SRM program, say due to war or political unpopularity, the consequent termination shock would hit the climate with the full complement of warming almost immediately. There'd be no time to adapt. Termination shock is often imagined as an outcome of war or fiscal strain, but that needn't be the case. It could be fully intentional. Andreas Malm has argued the tendency of side effects of SRM to rise over time suggests a tendency toward termination shock. Suppose SRM is implemented in the 2030s and carried on for a century. Malm imagined a future in which, after anyone with living memory of global warming is gone, mounting side effects lead policymakers to conclude the damage is worse than a few degrees. Any masked temperature (perhaps 4 to 5°C) could return inside of a decade.[44] Far sooner, it isn't hard to imagine one or two

42  Ed King, "Scientists Warn Earth Cooling Proposals Are No Climate 'Silver Bullet,'" Climate Home News, June 14, 2013.

43  Klein, *This Changes Everything*, 276.

44  Andreas Malm, "The Future Is the Termination Shock: On the Antinomies and Psychopathologies of Geoengineering. Part One," *Historical Materialism* 30, no. 4 (2022): 3–53.

dry monsoon seasons and the images of starved millions leading to termination demands at the UN.

We base this fantasy on one volcano from decades ago. It's easier to imagine dialing down the sunlight than the end of capitalism.

At the end of the litany of woes, the aesthetics are all wrong. Amid poisoned oceans and starvation, what lingers in my mind is the white sky. As every schoolchild learns, our blue sky results from the scattering of light by a particular mixture of particles overhead. Stratospheric aerosols would change wavelengths reaching the surface. At sunset on an SRM Engineered Planet, the sky is deeper red. At night, the stars are gone. In the daylight, the sky is a milky haze.

Perhaps, a few generations from now, children will stare up at the sky in disbelief. It cannot be true, they might think, what their grandparents remember of days when the skies were deep blue. Whether that child eats well or starves comes down to where they live under white skies.

# PART V

*Barbarism*

# 11

# The Greatest Migration

*Walter Benjamin asked me once in Paris during his emigration,*
*when I was still returning to Germany sporadically, whether there*
*were really enough torturers back there to carry out the orders of*
*the Nazis. There were enough.*

—Theodor Adorno

*A real-world interpenetration of apocalypse and utopia. Apocalypse*
*for those thousands who drowned on their own lungs. And for*
*the corporations, now reassured that the poor, unlike profit, were*
*indeed dispensable? An everyday utopia.*
    *This is another of the limitations of utopia: we live in utopia; it*
*just isn't ours.*

—China Miéville

One hundred and sixty set out from San Pedro Sula in Honduras.
A flier promoting the exodus depicts a lone man wearing a back-
pack, long sleeves, and hat to protect him from the elements. He
walks with arms outstretched into a red background. Above him
text reads, "No nos vamps torque queremos, nos expels la violence
y la pobreza" ("We do not leave because we want to, the violence
and poverty expel us").[1]

It is October 2019. Migrants answer the call on social media
from left activists to travel openly together. East through Guatemala and Mexico. Pivot north toward the United States. In little
more than a week, their ranks swell to as many as seven thousand.
Journalists dub it a "migrant caravan."

---

[1]  Azam Ahmed, Katie Rogers, and Jeff Ernst, "How the Migrant
Caravan Became a Trump Election Strategy," *New York Times*, August 24,
2018, nytimes.com.

Some interpret the movement as a critique of a troubled Honduran government, which it doubtless was at the start. As families from Honduras, Guatemala, Nicaragua, and El Salvador join, other observers suggest it is blowback from the coups, death squads, and genocides supported by the United States in the region over the last half century.

Increasingly, inhospitable Central American states of Guatemala, El Salvador, Honduras, and Nicaragua sit in a dry corridor vulnerable to irregular rainfall. The worst drought in four decades deprives a quarter of the population of nutrition. A Honduran coffee farmer, Antonio Lara, joins the caravan, because "coffee used to be worth something, but it's been seven years since there was a decent price."[2] In the warmer climate a fungus spreads and harms seventy percent of coffee farms in the region. Jesús Canan, a Ch'orti' Maya farmer, joins the caravan after his maize crop fails. "It didn't rain this year. Last year it didn't rain . . . My maize field didn't produce a thing." Desperation everywhere. "If we are going to die anyway," says the Honduran farmer Jorge Reyes in the anguish of hunger, "we might as well die trying to get to the United States."

It's unclear whether US conservative media intentionally recapitulate the UK politician Nigel Farage's 2016 migrant caravan poster aping Nazi-style propaganda, but amplify fright they do. While the caravan is still many thousands of kilometers south, Fox News runs B-roll footage of migrants amassed at a border (never mind the small text noting the border is Guatemala's). Daily garbage screams at geriatric white audiences about the enemy at the gates. Quickly, the panic spins up a tactic left dormant for two decades: blaming Muslim terrorists.

On October 16, President Trump tweets for the first time about the caravan. Two days later he inveighs, "Sadly, it looks like Mexico's Police and Military are unable to stop the Caravan heading to the Southern Border of the United States. Criminals and unknown Middle Easterners are mixed in." Right-wing media paint families as ISIS insurgents. On October 18, he punches again. "The assault on our country at our Southern Border, including the Criminal elements and DRUGS . . . All Democrats [sic] fault for weak laws!" By

2   Oliver Milman, Emily Holden, and David Agren, "The Unseen Driver Behind the Migrant Caravan: Climate Change," *The Guardian*, August 30, 2018, theguardian.com.

end of month, he warns of "Many Gang Members and some very bad people" among the crowd. "This is an invasion of our Country and our Military is waiting for you!"

Witness the future of denial in the resurrection of a long-dormant gimmick. After 9/11, the Department of Homeland Security was established alongside fabricated claims that al-Qaeda was pouring over the southern border. Now the president's mind, flickering in and out of timelines, resurrects the myth. Or perhaps his advisors mimic European xenophobia fusing immigration with Islamophobia. Whatever the chain of signifiers, Trump stammers about border guards capturing "people from the Middle East" and the Islamic State. He isn't alone.

On October 11, Guatemala's president Jimmy Morales claims they have captured and deported one hundred ISIS terrorists. Not to be outdone, White House Press Secretary Sarah Sanders soon lies that four thousand terrorists were captured by border patrol in a year.

Trump promotes an anti-Semitic video claiming to show that George Soros funds the caravan, "Soros" being a synonym for "Jewish money" on the right. Representative Matt Gaetz also blames the caravan on Jewish money. The radio host Michael Savage says the caravan signals "the end of America as we know it." One-time presidential candidate and media personality Pat Buchanan warns, "Yet far more critical to the future of our civilization is the ongoing invasion of the West from the Third World." Former Speaker of the House Newt Gingrich joins in, "If you were a terrorist and wanted to get in the United States and you saw 10,000 people trying to get into the United States, how unlikely is it that you might decide to join them?"

In the midst of this, on October 22, authorities arrest a Trump enthusiast named Cesar Sayoc. He is mailing pipe bombs to Soros along with prominent Democrats and public figures. On October 27, Robert Bowers, who convinces himself the Jews are scheming to replace his countrymen with immigrants, shoots seventeen people, murdering eleven at the Tree of Life Synagogue in Pittsburgh. A turbulence of bigotries is converted into terror.

The caravan reaches the border. They request asylum. The United States forces Mexico to settle those who don't return home. Their stories are harsh and tragic.

## The Beginning of the Greatest Migration

Theodor Adorno said that every debate about the ideals of education is trivial compare to a new categorical imperative imposed by the Shoah: never again. We must arrange all our thoughts and actions so as to prevent another relapse into barbarism. In that spirit, I tell every student we study the humanities because we live at the start of the greatest migration there ever has been or ever will be. It's the real reason for this book. The responsibility to understand difference and learn from the failures of those who went before us is a heavy burden.

Seven in ten Americans believe global warming is happening, but only 57 percent believe it is caused by human activities. Six in ten say it will harm people in the United States, and seven in ten believe it will harm developing nations. There's a massive mismatch between the number of people affected, everyone, and the number of people who will be seen as victims requiring assistance.[3] None of what's coming will feel like an apocalypse but, instead, more of the already unbearable same. In the North, the horrors will be absorbed as unfortunate entertainment ("Did you see what happened? The poor souls!") but inconsequential for viewers insulated from the worst. For a while.

In America, the Great Migration refers to a period in the first half of the twentieth century, when African Americans relocated from the Jim Crow South to the Midwest and North. Cities grew, economies were reconfigured, and neighborhoods were integrated. Self-determination competed against the Great Depression and white backlash in these transformative decades. Already underway, the Greatest Migration will displace communities, erode ecosystems, disfigure and supplant old ways, and remake societies at orders of magnitude for which there's little precedent.

The right will be ready. The GOP is the most dangerous organization in the world, but it is not unique in climate denial or commitment to the destruction of organized civilization. In *White Skin, Black Fuel*, a superb catalog of denial in the European right, Andreas Malm and the Zetkin Collective identified a curious trope: "Every time a European far-right party denies or downplays climate

---

3   Jennifer Marlon et al., "Yale Climate Opinion Maps 2021," Yale Program on Climate Change Communication, February 23, 2022, climate communication.yale.edu.

change, it makes a statement about immigration."[4] Recall Freud's kettle logic: climate change is happening, or it's not, or who can say? Focus instead, says the right, on immigrants, Muslims, Syrians, and Africans.

Fossil fascism grows on the right. The bourgeois denialist ISA—the fossil fuel companies, conservative think tanks, conservative media—has always relied on reactionary proletarians to believe its counterfeit sciences blaming sun cycles or cooling myths. But as old climate denial bumps up against the limits of credulity, the owner class now tolerates open xenophobia in the reactionary base as a motivational tool sabotaging mitigation.

Anti-Semitism circulates openly in right-wing climate denial, where tropes might as well be lifted from the *Protocols of the Elders of Zion* with a little fine-tuning (Soros funds nefarious schemes, global elite control, and so on). At the same time, the figure of the Jew is supplanted with more acceptable xenophobias against Africans, Syrians, or Muslims. European and North American xenophobias exchange notes and adjust for regional particularities. Maybe that's what happened when, in 2018, Fox News and Trump repeated Farage's 2016 migrant caravan drumbeat. Farage tapped European racism against African Muslims. But the US right codes Muslims specifically as Arabs, not Africans, and it detests both Muslims and African Americans, but separately. So when the US right saw a migrant caravan it resurrected an old myth of al-Qaeda operatives pouring over the United States–Mexico border.

Excuses for violence don't need coherence. Affect is enough. Otherness is enough. Like a conversion symptom, denial turns into barbarism. This chapter casts a wide net of loosely connected vignettes of barbarism to think about the horizon.

## Border Externalization

Late in late winter 2020, Greece announced plans to erect nets in the Aegean Sea. Too many Syrians, they decided, over seventy thousand per year. Refugees fled a civil war resulting from displaced people concentrated in hot cities during a drought. Turkey was aggravated over Syrian refugees on its soil and happily passed them to a neighbor, where the state built new island camps.

---

4 Malm et al., *White Skin, Black Fuel*, 38.

Backlash ensued as locals staged protests to demand, "We want our islands back." So instead of camps, erect nets.[5]

The nets would stretch several kilometers near the island of Lesbos. They would rise half a meter out of the water with flashing lights to either catch or drown Syrians. The left party Syriza called the project "a disgrace and an insult to humanity." The Greek defense minister deflected, claiming the nets were meant to catch smugglers. Amnesty International condemned the "alarming escalation" and asked whether authorities would respond to distress signals from trapped rafts.

The Greek government insisted they'd test nets for safety. The European Commission said the nets were legal so long as they didn't prohibit migrants from seeking asylum. The Greek migration minister minced fewer words, defiantly warning the net "sends out the message that we are not a place where anything goes and that we're taking all necessary measures to protect the borders." So much handwringing and dancing around what all can plainly see. These nets would be floating kill zones. The crisis emerged from a warmer climate, but nets don't discriminate.

For as long as borders have existed, back to stone walls and neolithic fencing, humans and animals have always moved around borders. Walls and fences are one technology of borders, surely the most visible, but border technologies also include visas and passports, which distribute rights unevenly and unpredictably. For example, when the Trump administration enacted its Muslim ban, the legal status of US-bound visa holders switched mid-flight, while green card holders mostly remained protected. Borders do not keep people out. What borders do is confer and revoke status. We tax goods at the border. Humans turn "illegal" at a border.[6] As Catherine Dauvergne said, "Illegal migration is a product of migration law. Without legal prohibition, there is no illegality."[7]

Borders literally map out possible movements. Compare the freedom to roam within the Schengen Area to the anxiety that

5  "Greek Plan to Stop Migrant Boats with Float Barriers," *BBC News*, January 30, 2020, bbc.com; and Niki Kitsantonis, "Greece's Answer to Migrants, a Floating Barrier, Is Called a 'Disgrace,'" *New York Times*, February 1, 2020, nytimes.com.

6  See Thomas Nail, *Theory of the Border* (Oxford: Oxford University Press, 2016).

7  Catherine Dauvergne as quoted in Harsha Walia, *Border and Rule* (Chicago: Haymarket Books, 2021), 5.

migrants from just outside the Schengen Area might enter and fan out. This fear lurked across the Brexit debate, though the UK isn't part of the Schengen Area. When the Syrian migrant crisis picked up in 2015, images of drowned families vividly testified to brutally uneven rights distributed at borders.

Borders are fickle. Now they're externalized. Concentration camps and border patrol raids draw attention, but power acts on migrants or refugees long before they reach a host nation. In her marvelously documented book *Border and Rule*, the immigrant rights activist Harsha Walia defined border externalization as webs of "interdiction, offshore detention, safe third country agreements, and outsourcing of border control to third countries."[8] While signatories to the Universal Declaration of Human Rights and UN Refugee Convention are obliged to grant asylum to persecuted people and protect refugees, the Global North views refugees as undesirables who should stay in their nation of origin. If they leave their homelands, migrants must then seek asylum in a neighbor country in Central America or Africa.

Externalization entangles direct foreign aid and private-public partnerships. Examples include the Overseas Private Investment Corporation (now US International Development Finance Corporation) as well as the US–funded Joint Border Intelligence Group, which works with the Department of Homeland Security to police borders in the Northern Triangle under pretenses of rooting out gangs, drugs, and human trafficking. The crackdown on Central American migration is now largely outsourced south of our border but paid for with US dollars. Emphasis on narco-trafficking justifies paramilitarization, and drumming up human trafficking paints racism as a humanitarian mission. Meanwhile migrants disappear. In all likelihood, many are massacred. Eighty percent of the women report rape by captors.[9] In a recent three-year period, Mexican patrols apprehended a half million Central American migrants, and seventy thousand migrants disappeared. Some activists call it a "migrant holocaust." It is under these barbaric conditions that the United States demands Mexico get tough.

DHS isn't cagey about externalization. After settling a Mexico-Guatemala-Belize border agreement, a DHS official declared, "The Guatemalan border with Chiapas is now our southern

8   Walia, *Border and Rule*, 87.
9   Ibid., 91.

border."[10] So, back in 2018, when Fox News showed deceptive footage of a migrant caravan at Mexico's southern border but left the unscrutinizing viewers with the impression that the migrants were already at "the border," there was a bit of truth, if only truth meant to deceive. US officials do indeed view all of Central America as a border buffer zone. While trade policy, most notoriously NAFTA, impoverishes communities and drives people north, the so-called War on Drugs justifies repression of southern neighbors.

## Fortress Europe

On the northern coast of Africa in May 2021, men, women, and children struggled around a barrier separating Morocco from the autonomous Spanish city Ceuta. Morocco loosened border controls over a spat with Spain. Eight thousand crossed in a couple of days.

Spanish authorities responded aggressively. Armored personnel carriers rolled in as soldiers wielding shields and batons chased migrants. Spain claimed its patrols pushed more than half the crowd back to Morocco.

Viral scenes spread online. On an otherwise nondescript beach, a jetty extended into the water. Atop the rocky barrier, a transparent fence rose another six meters, barbed wire and motion sensors adorned the top of the barrier. Sensors weren't needed, though, when thousands of desperate people rushed the guards. Spanish soldiers grew angry at Moroccans swimming around the barrier from their side to Ceuta.

Migrants sidestepped carefully along the Moroccan side of the jetty before rotating around the wall to the Ceuta side. One mustn't lose one's footing in such a situation rife with desperation as bodies crowd above sharp rocks in the surf below. Spanish soldiers swung batons as migrants rounded the corner to the Ceuta side and pushed them into the water. They quite literally pushed migrants into the sea.

Since a 1992 agreement between Spain and Morocco, the latter has been compelled to readmit deportees caught in autonomous cities like Ceuta and Melilla along with whoever crosses successfully to the Iberian Peninsula. As with North American governments,

---

10   Ibid., 90.

European border externalization tries to return migrants to an origin country or halt them in transit.

Fortress Europe is the colloquial name for a sophisticated regime of agencies and technologies guarding the continent from migrants. Traditional walls and razor wire are one component, but big budgets integrate comprehensive tools: biometric data, "smart border" systems with artificial intelligence to screen migrants, expanded patrols for the EU border agency Frontex, drones for the European Border Surveillance System, readmission requirements written into aid and trade agreements, EU soldiers training a new counterterrorism and anti-immigration force from African states called the G5 Sahel Cross-Border Joint Force, and surveillance, military equipment, and prisons paid for by the EU Emergency Trust Fund for Africa.[11] The results? Migration from Africa to Europe decreases, but fatalities continue to grow, as migrants are redirected to more dangerous paths.

The International Organization for Immigration estimated 33,761 people died or went missing in the Mediterranean Sea between 2000 and 2017. The death toll pales in comparison to lethal routes across the Sahara, which claim twice as many lives.

There are three routes to Europe. We have already examined the first, the western route from Morocco up to Ceuta and Melilla. Second, a central route runs from Libya or Tunisia to Italy. Third, an eastern route runs from Turkey to Greece, through the Aegean Sea. In response to more patrols in one area, migrants shift vectors. Walia argued the EU bears responsibility for driving families onto progressively more dangerous paths. "Though the media characterizes them as 'new migrant routes'—as though migrants are merrily Columbusing their way around—these paths are actually 'EU-organized channels.'"[12]

The central route is a particularly brutal example. In compliance with European demands, Algerian officials deflect migrants into neighboring Niger, where they are redirected further into Chad or Sudan before a pivot to Libya and pressing north to the sea. The wild path claims many lives, all according to plan or at least a predictable outcome. Those who cross successfully might find themselves in camps, such as the Moria refugee camp on the island of Lesbos, which holds tens of thousands crammed into tents.

---

11  Ibid., 108–9.
12  Ibid., 110.

Humanitarian laws provide a pretext for brutality. Laws targeting human trafficking, or "modern-day slavery," are used to crack down on migration. The same is true in the United States and EU for laws on "trafficking and smuggling." In a perverse twist of priorities, some refugees themselves face prison if convicted on charges of "facilitation of irregular entry and stay" for help with blankets and food. Yes, laws punish aid to refugees as "facilitation" of human trafficking. An analogous problem exists in the US, where heroes who stow water in the deserts near the United States–Mexico border get criminal charges.

It isn't hard to imagine Fortress Europe adapting to the Changes. If aid and trade deals are already tied to migration controls, won't adaptation aid fund border repression? Europe needn't build so many walls on its coasts if it can bribe other governments to police its neighbors, pushing migrants back at the borders or sending them to wander in the Sahara.

## Drones

Anytime in the so-called War on Terror, drones fling missiles carelessly. Regions hit lie on the aridity line where barely enough rain falls for crops. The southern aridity line runs across the African Sahel, while another tracks along the northern edge of Africa before winding through the Middle East.

Thanks to externalization, refugees are slowed down and remain in Africa. Heat-stressed and drought-stricken areas are packed tighter still. Some rebel. Those who do are labeled terrorists and slated for termination. All for lack of rainfall on an aridity line that, thanks to warming, moves year by year.

I first encountered the jarring map of drone-strike locations following the aridity line winding through Africa and the Middle East in Naomi Klein's *On Fire*. Her inspiration was Eyal Weizman's forensic architecture. "Recent conflicts along the aridity line erupted because of these environmental transformations," Weizman wrote. "Droughts led to increased competition over shrinking resources and aggravated the consequences of civil strife. The land degradation brought about by conflict further aggravated desertification."[13]

---

13  Graph adapted from Naomi Klein, *On Fire* (New York: Simon & Schuster, 2019), 162; and Naomi Klein, "Let Them Drown," Positions

The aridity line moves through Syria at the border city of Darra, where farmers' protests born of drought lit up the Syrian civil war in 2011. The line passes through Raqqah, the base for the Islamic State as well as areas of fierce fighting in Afghanistan and Pakistan. It traces along conflict zones at the northern and southern edges of the Sahara. Per Weizman: "Plotting the location of western drone strikes on meteorological maps demonstrates another astounding coincidence: many of these attacks—from South Waziristan through northern Yemen, Somalia, Mali, Iraq, Gaza, and Libya—are directly on or close to the 200 mm aridity line."[14]

Climate change augments preexisting tension. It's a consequence of industrialization, but Weizman reminded us, "The climate has always been a project for colonial powers, which have continuously acted to engineer it." Plot the advance of desertification, and one might see where drones will soon lurk. Meanwhile, Westerners nervously whisper of approaching water wars, as if we are not already engaged.

If mechanized, routinized violence distributes along the aridity line, we are likely to see the same technologies return home. At our borders, maybe within them, we should expect to see automated weapons platforms integrated into Fortress Europe or Fortress Americas. Don't take it as doomsaying. Israel, for instance, already polices its borders with robotic killers.

In Gaza, which is often called the world's largest open-air prison, two million people are crammed inside 365 square kilometers. The average age is eighteen, and two-thirds are under twenty-five years of age. The young population gets access to water a few hours per day, and 80 percent depend on international aid. Unemployment runs as high as 70 percent. Israel polices the border with a sprawling panoply of semi-autonomous and guided weapons platforms. This was Gaza's status quo before Israel's genocidal assault, which began in October 2023 and continues as of this writing to show how states liquidate unwanted people

One platform is the "robo-sniper." Perched atop the Gazan strip border fence, machine guns can detect human movement and fire from a kilometer and a half away, quite far into a strip of land only six to twelve kilometers across. Turrets can be equipped with

---

Politics, accessed May 13, 2022, positionspolitics.org. Quote from Eyal Weizman, *Erasure* (Göttingen: Steidl, 2015), 9–10.

14  Ibid., 10.

rockets to eliminate vehicles or humans. Multiple guns can be trained on one individual.

Next, the suicide drone can hover in the air, loitering until given an order to attack. The drone is theoretically autonomous but can also be leashed to a "man in the loop."

The latest upgrades to semi-autonomous drones include several rovers. The smaller one-and-a-half-tonne Jaguar is a six-wheeled platform supporting a frame laden with high-resolution cameras, headlights, and a machine gun. It can fire on humans detected over a kilometer away. Though operators claim the Jaguar is not fully autonomous in its border patrol, its program can fire without human input. When asked about the moral problems posed by tablet-operated hunter-killer robots that could go fully autonomous, a developer of a larger rover called the REX MKII deflected and said, "It is a decision of the user today." Israel deploys large numbers of smaller arial drones for surveillance. The IDF literally calls these "drone swarms" where fifty drones fly together with a smarter drone guiding others down into this or that street to surveil Palestinians. And the Jaguar can be armed with less-lethal weapons like tear gas or rubber bullets in order to perform semi-autonomous crowd control.[15] Perhaps drones will start showing up at protests near you.

In spring 2021, Israel announced the integration of these platforms. Weeks later, Israel assaulted Gaza, the so-called mowing the grass procedure in which every so often it inflicts overwhelming force for a few weeks. IDF declared it the "first AI war."[16] Of course, while the United States funds Israel's military, Israeli operators train American police.

When we speak of the possibility of sentry drones guarding borders and brutalizing migrants as the evacuations, displacements, and full-scale exoduses ramp up—to say nothing of crowd control for desperate climate protests and direct actions within our walled states—this is not some hyperbolic, dystopian possibility. It would be the continuation of now.

---

15 Sébastien Roblin, "Israel Is Sending Robots with Machine Guns to the Gaza Border," The Daily Beast, January 25, 2021, thedailybeast.com; and "The Future of Defending Israel 'Jaguar': The IDF's Newest, Most Advanced Robot," Israel Defense Forces, April 27, 2021, idf.il.

16 Judah Ari Gross, "In Apparent World First, IDF Deployed Drone Swarms in Gaza Fighting," Times of Israel, July 10, 2021, timesofisrael.com.

## The Largest Camp

Bangladesh's Kutupalong camp, the largest refugee camp in the world, is critically full with Rohingya driven out of Myanmar. Half a million undocumented Rohingya settled here even before the Myanmar military's genocide displaced another seven hundred thousand across the border in 2017. Cyclones and floods threaten the camp. The government has already relocated a substantial portion of the population from the current camp in Cox's Bazar to an island called Bhasan Char.

They are all terribly vulnerable to sea level rise. A meter rise would submerge 20 percent of Bangladesh. Projections show much of Cox's Bazar flooded in three decades. The average elevation of Cox's Bazar is three meters above sea level. Bhasan Char's is only two.[17]

"They lured us with the promise of good health care facilities, good food," recalls one Rohingya woman. "But after coming here I find that we are not given proper health care or medicines."

The low island Bhasan Char is barely twenty years old, formed from silt from the Meghna River in 1999. The island's shorelines shift dramatically, expand and contract, change shape. Time-lapse photography appears to show the entire island hop up and down in latitude. On this shifting sand the government moves human beings over the protests of United Nations inspectors who aren't permitted to check safety. Children receive no formal education and die for lack of medications widely available on the mainland even in the Cox's Bazar camp. Men and women complain of a lack of food or gas for cooking. People turn to fishing. It's not enough. Water is unreliable. Parents are prohibited from taking families back to the mainland camp. One man, Yusuf Ali, said his daughters on the island felt it was like "an island jail in the middle of the sea." Many Rohingya express a desire to return home to Myanmar, trying their chances with genocide.

Diseases, fires, and assaults are constant in the camps, the horrors predictable when people are left without resources. Bangladeshi security forces regularly kill or abduct refugees. Or refugees are

---

17   Meenakshi Ganguly, ed., "An Island Jail in the Middle of the Sea," Human Rights Watch, June 7, 2021, hrw.org; and Moazzem Hossain and Swaminathan Natarajan, "The Rohingya Refugees Trapped on a Remote Island Miles from Land," *BBC News*, May 29, 2021, bbc.com.

picked up by the navy in the ocean and beaten before transfer to the camps. In a past response to protest, the Bangladeshi authorities shut off internet access in Cox's Bazar, creating a dangerous vacuum of knowledge during the Covid-19 pandemic.

Thailand doesn't want more Rohingya. Neither does Malaysia. India grows impatient with the refugees picked up by their coast guard and threatens to deport forty thousand Rohingya, but to where? Bangladesh won't return them to Myanmar, where they would be persecuted. So they stay on Bhasan Char.

Still, they won't be welcome forever, not even in the drowning camp. "Has Bangladesh been given the global contract and responsibility to take and rehabilitate all the Rohingya or boat people of the world?" asks Foreign Minister A. K. Abdul Momen. "No, not at all."

Powerful cyclones pummel the Bay of Bengal each year. Ferries to and from the island take three to five hours and will not be reliable options for evacuation. Helicopters are grounded in cyclones. Storm surge in a cyclone frequently exceeds five meters. Housing on the island is propped up on stilts about one meter off the ground.

Already at high tide in a storm, the island could be submerged—even before the seas rise and the cyclones supercharge. Authorities will move hundreds of thousands of refugees to an island that will be drowned.[18]

The conversion symptoms proliferate.

## The Climate Migrants

"And this brings me to the challenge I want to focus on today," President Obama pivoted in a commencement address at the Coast Guard Academy. "That's the urgent need to combat and adapt to climate change."[19]

He drew upon a quote to say the military must adjust: "The

---

18   "Rohingya Refugees Describe the Conditions on Bhasan Char," YouTube video, Human Rights Watch, 2021; "Why Are Rohingya Refugees Attempting an Escape from Bhasan Char?" YouTube video; WION Ground Report, 2021; and Ganguly, "An Island Jail in the Middle of the Sea."

19   Barack Obama, "Remarks by the President at the United States Coast Guard Academy Commencement," White House, May 20, 2015, obamawhitehouse.archives.gov.

pessimist complains about the wind; the optimist expects it to change; the realist adjusts the sails." He warned of instability and invoked the climate migrant.

"Globally, we could see a rise in climate change refugees. And I guarantee you the Coast Guard will have to respond." Did he mean to help or hurt? How did he contextualize the Coast Guard's responsibility?

He spoke about mass migration and reminded graduates the Pentagon called climate change a "threat multiplier." He pointed to a drought in Nigeria exploited by Boko Haram, crop failures driving unrest in Syria, and Super Typhoon Haiyan, which killed more than ten thousand. They'll conduct more humanitarian missions, Obama insisted, but watch for threats to homeland security and infrastructure.

Tying climate concerns ever closer to the military budget, he moralized, "Politicians who say they care about military readiness ought to care about this, as well." He spoke of greening the military with biofuels. Ominous nonsense.

"We need you in the Caribbean and Central America, interdicting drugs before they reach our streets and damage our kids." In the Changes, the War on Drugs is a pretext for repression. "We need you in the Middle East; in the Gulf; alongside our Navy; in places like West Africa, where you helped keep the ports open so that the world could fight a deadly disease." In the Changes, open ports and disease scares are excuses for military adventurism. "We need you in the Asia Pacific, to help our partners train their own coast guards to uphold maritime security and freedom of navigation in waters vital to our global economy." In the Changes, aid for allies means keeping the ports open and the treats flowing to the North. In a buffet of concern, Obama glided between the poor migrant and policing subaltern populations.

Climate refugees. Environmental migrants. Climate change migrants. Each term circumscribes humans in rhetorical slots and designates who deserves help, who doesn't, and why. In Walia's words, "Refugees are often characterized as fleeing persecution in search of safety, while migrants are depicted as 'bogus refugees' moving for economic reasons and hence undeserving of protection." Migrants are often seen as upwardly mobile, while refugees must be tended and watched. Sprawling legal conventions could feel overwhelming even if we had a clear definition for a climate migrant, but we don't.

The 1951 UN Convention Relating to the Status of Refugees defined a refugee as having a "well-founded fear of being persecuted for reasons of race, religion, nationality, membership of a particular social group or political opinion." No environmental protections exist. The refugee "is outside the country of his nationality and is unable or, owing to such fear, is unwilling to avail himself of the protection of that country; or who, not having a nationality and being outside the country of his former habitual residence as a result of such events, is unable or, owing to such fear, is unwilling to return to it."[20] It's about fear of violent people not an inability to return to an island now underwater.

According to the International Organization for Immigration, terms like "environmental migrant" or "climate change refugee" are misleading and undermine extant legal protections for refugees. IOM uses a nonbinding definition for environmental migrants, based on these qualifications:

> Environmental migrants are not only those displaced by extreme environmental events but also those whose migration is triggered by deteriorating environmental conditions;
>
> environmentally-induced movement can take place within as well as across international borders;
>
> it can be both short and long term; and
>
> population movements triggered by environmental factors can be forced as well as a matter of choice.[21]

The Paris Agreement contained a single reference to migrants and none to refugees. It spends exactly as much text on climate refugees as on fossil fuels. The one mention:

> *Acknowledging* that climate change is a common concern of human-kind, Parties should, when taking action to address climate change, respect, promote and consider their respective obligations on human rights, the right to health, the rights of indigenous peoples, local communities, migrants, children, persons with disabilities and people in vulnerable situations and the right to development, as well

---

20 Article 1 of "Convention and Protocol Relating to the Status of Refugees," United Nations High Commissioner for Refugees, 1951.

21 Definitions from "Definitional Issues," International Organization for Migration, accessed May 5, 2022, iom.int.

as gender equality, empowerment of women and intergenerational equity.[22]

Blink and you miss it. How can there be no legal box for climate migrant while people are displaced?

IOM ticked off a few challenges for a new category.[23] First, currently, migration is mostly internal and seeks domestic protections, so they are largely invisible. Second, migration may be forced or voluntary. Third, a new refugee convention risks scrubbing existing protections. Forth, adaptation (even legal adaptation) always threatens to displace focus on mitigation. Fifth, migration pathways already exist. To me, these explanations range from reasonable to strained, but one final reason stands above the rest: it's difficult to isolate environmental reasons as a singular cause for relocation.

Climate migrants might not even see themselves as such. When surveying participants in the 2018 Central American migrant caravan, IOM found migrants joined for reasons such as "seeking better conditions," "violence/insecurity," and "family reunification." Refugees or migrants might not themselves perceive the ecological reasons for their physical and economic insecurity. If that sounds wrong, ask yourself how many wealthy Californians who move to less fire-prone regions perceive themselves as climate refugees. Climate change is such a daunting existential crisis because it exacerbates prior problems. The Norwegian Refugee Council says disasters displace three to ten times more than conflict and war.[24] If conflicts are driven by scarce resources, at what point does it become impossible to isolate the relationship of climate change to migration?

Walia criticized IOM as a pillar of immigration management. From its founding in 1951, the same year as the UN Refugee Convention, IOM "was initially propped up by the US to provide services and logistics to contain global migration in line with the Cold War agenda." Today, IOM facilitates "assisted voluntary returns," those so-called self-deportations, and, broadly speaking,

---

22   Preamble to "Paris Agreement," United Nations Framework Convention on Climate Change, 2015.

23   Dina Ionesco, "Let's Talk About Climate Migrants, Not Climate Refugees—United Nations Sustainable Development," United Nations, June 6, 2019, un.org.

24   Baher Kamal, "Climate Migrants Might Reach One Billion by 2050," ReliefWeb, 2017, reliefweb.int.

migrant aid organizations "created a slew of legally distinct categories and a hierarchy of rights in order to manage, divide, and
control people on the move."[25]

Capitalist climate governance will likely treat these vulnerable
people, stateless and without legal protection, as roughly as it treats
forgotten refugees today. Fascist and reactionary governments will
turn clumsier with concentration camps. Doubtless, some billionaires will try to backwards engineer a mid-twenty-first-century
version of early twentieth-century company towns, turning migration into a source of exploitable labor or slavery by another name.
None of these paths deviate from already-existing capitalism.

How many people will move? Estimates cited by the UN range
from as little as a few tens of millions to as many as a billion by
2050. The Biden administration projected internal displacements
in sub-Saharan Africa, South Asia, and Latin America at 143
million people, or 3 percent of the population in those areas by
2050.[26] Oxfam now estimates climate-fueled disasters internally
displace twenty million per year—a person every second or two,
a five-fold increase over the last decade. People were three times
more likely to be displaced by cyclones, floods, and wildfires than
by conflict. Oxfam's term "climate-fueled disasters" underscored
the challenge of designating climate change as a neatly defined
cause, however clearly present. If we were to extend Oxfam's
displacement rate (which will, in reality, surely grow worse) over
three decades, we'd reach 600 million displaced. Focusing on
internal migration alone, that number is already on the higher end
of midcentury forecasts.[27]

India and sub-Saharan Africa will suffer the worst. One study
found that, under SSP5-8.5, nearly a billion Indians and half a
billion Nigerians will be displaced by end of century, whereas
China, Russia, and the United States will absorb migrants. In
SSP2-4.5, the migration numbers are about half as bad, but in the
unabated scenario SSP5-8.5 a staggering 3.8 billion people will
move by end of century.[28]

25  Walia, *Border and Rule*, 86–7.

26  "Report on the Impact of Climate Change on Migration," White
House, August 2021, whitehouse.gov.

27  "Climate Fueled Disasters Number One Driver of Internal Displacement Globally Forcing More than 20 Million People a Year from
Their Homes," Oxfam International, December 2, 2019, oxfam.org.

28  M. Chen and K. Caldeira, "Climate Change as an Incentive for

Recent IPCC estimates say 250 million will face water stress in Africa by 2030, displacing 700 million. Sub-Saharan floods will displace 2.7 million in any given year. Sixty million more will face malaria risk.[29]

Who are the migrants? "Climate change is modifying existing migration trends, and people are maybe migrating more frequently or to farther destinations than they previously were," said Julia Blocher, a researcher with the Potsdam Institute for Climate Impact Research. "We see less evidence of new migration flows, except of course in the case of natural hazard-induced disasters." Rather than a mass exodus, "it's much more likely to be people who are living agricultural livelihoods making a difficult decision to migrate for wage employment . . . they may not have made that decision otherwise, or perhaps moving more frequently." Migration often means men of working age relocating from rural to urban areas for employment or women moving for education or marriage, and migrants tend to be young. They face terrible risks of sexual violence, wage theft, and other forms of exploitation.[30]

The shift from rural, agricultural livelihoods to urban work means the migrant is at a juncture of mitigation and adaptation projects. "A lot of cities across the world are not necessarily prepared for the possibility of not only a much higher population than they have currently but at the same time the impacts of climate change," Blocher said. If risk assessments measure exposure of current populations without unpredictable influxes, even reasonable forecasts of climate impacts won't prepare cities for risks of ballooning urban population.

The pandemic cost younger people crucial years of career development. "What I think we'll see as a result is more and more people either migrating for wage labor that they might not have otherwise done, or, on the other side of the same coin, some reverse migration." Blocher drew upon data in Tanzania, where people moved out of cities during the pandemic to feel safer in rural areas and sought work among family. This reverses the common

Future Human Migration," *Earth System Dynamics* 11, no. 4 (2020): 875–83.

29  Intergovernmental Panel on Climate Change, *Sixth Assessment Report*, Working Group II (2022), 9–148.

30  Julia Blocher, "COP26: Why Migration Policy Is Critical to Climate Adaptation," interview by Ty Benefiel and Brock Benefiel, *The Climate Pod*, November 8, 2021, podcasts.apple.com.

practice of a laborer relocating temporarily away from family to work in a city. Additionally, migrants who move internationally through existing immigration frameworks tend to be the wealthiest and most educated in their communities, not the most marginalized.

If migrantions remain mostly internal, there are limits on what international rules can do. Foreign aid can be a tool of climate colonialism and border externalization. A "climate migrant" status will likely be integrated into preexisting regimes of racialized exploitation and apartheid. What we have, then, is vulnerable people facing institutional resistance to classification, their number ultimately uncountable yet threatened everywhere with barbarism.

## Exploitation in Export Processing Zones

Much like the coal company towns of old Appalachia, where guards enforced order on workers often paid in company currency and dependent on company homes, we now see the company region. We should keep in mind what Frantz Fanon said about Marxist analysis in colonies, to which we should at the very least add any region with a history of enslavement or apartheid. "In the colonies the economic substructure is also a super-structure. The cause is the consequence; you are rich because you are white, you are white because you are rich."

Fanon expanded Marx's metaphor of a legal and political superstructure holding together a society built upon a base of capitalist material relations. He argued that class analysis, after colonization, must flex to include how race is coded into the legal and political superstructure. The wealthiest 10 percent are responsible for fully half of cumulative emissions in the last three decades, but they won't be interned in the camps. In the Changes those who caused the least harm will be harmed most.

Export processing zones (EPZs) are legally circumscribed areas with alien rules facilitating capital. The World Bank describes them as specialized "in manufacturing for export to more advanced market economies by offering exporters duty-free imports, a favorable business environment, few regulatory restrictions, and a minimum of red tape."[31] Sometimes called "free trade zones," EPZs, according to Walia's definition, are "internal bordering regimes

---

31 "Export Processing Zones," The World Bank, 1992, worldbank.org.

of free capital flows and wage suppression—a capitalist haven of neoliberalism."[32] Garment factories are a common reason for EPZs. Ninety percent of workers in these zones are women. Depending on the definition, these types of zones are at least a half century old. But these areas, which are, in a very real sense, extranational bordered areas or company-states within states, aren't understood by the public.

Another model is the Honduran Economic Development Zone (ZEDE), literally called *ciudades modelos*, "model cities." ZEDEs are subject to Honduran military authority, but they are, as the immigration researcher Todd Miller explains,

> autonomous zones run by a technical secretary appointed by a committee that is in many ways separate from the sovereign country from which it was carved . . . zones each have their own laws and judicial systems, and their own government, serving the principles of free-market capitalism.[33]

Overseeing the creation of these zones stood a committee of right-wingers including Grover Norquist and Reagan speechwriter Mark Klugmann, who predicted "Central America could soon become—as southern China has been—the fastest-growing economic region in the world."[34]

Whatever the future models of these extrajudicial, extranational zones dedicated to surplus exploitation, the pattern isn't so different from the Chicago school villains who cheered Pinochet's coup as a pathway to a free-market fun zone. But capital needn't depend on villains. Capitalism incentivizes surplus exploitation and the extraction of natural resources, along with the leveraging and financialization of those resources. It reaps where it does not sow. What this means for our imagination of the Changes is unclear, but we can hazard a guess.

In recent years, with the number of EPZs on the one hand and migrant concentration camps on the other, no amount of undignified exploitation is off limits. Company towns and free-market regions devoid of labor law could prove to be crucial avenues of exploitation in the Global South, propping up a semblance of normalcy in

---

32  Walia, *Border and Rule*, 63.

33  Todd Miller, *Storming the Wall* (San Francisco: City Lights, 2017), 87.

34  Mark Klugmann in ibid.

the Global North. More speculatively, it's not impossible to imagine entire indebted, resource-stressed nations allowing themselves to be practically bought out by a behemoth like Amazon.

Perhaps we will see new hybrid migrant camps that are also company towns, or perhaps we will see entire Indigenous regions sold off to investment corporations as carbon offset reservations, which hedge funds will then incorporate into sophisticated financial instruments. Perhaps the difference between a corporation running private prisons for migrants and foreign factories in EPZs will cease to be a meaningful difference.

## Where Will They Go?

We previously considered the notorious difficulty of projecting climate lag on economies. Numbers range from the optimistic, barely more than 1 percent GDP loss for every 1°C increase, to the fantastic.[35] Oxford Economics forecasted 21 percent GDP lag at 3°C by 2100. They found India could lose as much as 90 percent.[36] Another study projected a 70 percent drop in the ten most affected nations by end of century, with Sudan most affected at 32 percent GDP reduction by midcentury and up to 84 percent by end of century.[37]

In Africa, the IPCC says GDP will drop 13.6 percent due to warming.[38] Impacts are severe in the Central African Republic, the Democratic Republic of Congo, and Zimbabwe. Decline in rainfall since 1960 has created a GDP gap between Africa and the rest of the developing world that would be 15–40 percent smaller without warming.[39]

By comparison, in the Great Depression, GDP dropped around 30 percent. The Covid-19 pandemic cost less than 7 percent. The

---

35  Solomon Hsiang et al., "Estimating Economic Damage from Climate Change in the United States," *Science* 356, no. 6345 (June 30, 2017): 1362–9.

36  James Nixon, "Modelling the Economic Cost of Global Warming" (Oxford Economics, July 6, 2020).

37  Marlowe Hood, "Climate on Track to Devastate World's Poorest Economies," Phys.org, November 8, 2021, phys.org.

38  Intergovernmental Panel on Climate Change, *Sixth Assessment Report*, Working Group II, 2022, 9–148.

39  Ibid., 9–130.

Changes will predominantly hit the developing world, at least for a while, but the North will not escape.

Famines will hit more frequently. As we saw in chapter 4, in the most optimistic numbers I've seen, crop yields drop 3.1–7.4 percent per degree of warming.[40] Several billion people already spend most of their income on food. What will happen when economic lag joins faltering crops and heat stress? Today only 1 percent of the global population lives in a barely livable hot zone. By 2070, that number could rise to one-fifth.[41] Will that mean a mass exodus, a clearing out of countries? What will it mean for borders and barbarism, and will your descendants even notice? Where will people go, and will they be met with welcome aid or sentry drones?

40  Aton, "For Crop Harvests."
41  Abrahm Lustgarten, "The Great Climate Migration Has Begun," *New York Times*, July 23, 2020, nytimes.com.

# 12

# The North Will Not
# Escape This Storm

*Fascist propaganda attacks bogies rather than real opponents, that is to say, it builds up an imagery of the Jew, or of the Communist, and tears it to pieces, without caring much how this imagery is related to reality.*

—Theodor Adorno

*The future is already here—it's just not very evenly distributed.*

—William Gibson

In the first summer of the pandemic, American streets erupt in protest after the killing of George Floyd, an unarmed Black man pinned to the ground and murdered by police. Protests in Portland carry on longer than in most cities. Demands for justice are met with tear gas.

Rumors of mysterious abductions circulate. The president threatens to quell uprisings with federal troops, but in fact they are already here. The US Marshals are present. The Border Patrol Tactical Unit (BORTAC) has been here for weeks. Secretive patrols sweep the streets.

The Department of Homeland Security's acting secretary will condemn demonstrations outside the courthouse and criticize local officials for failing to protect the city, which, he says, has been "under siege for 47 straight days by a violent mob." Take appropriate action, he demands. "DHS will not abdicate its solemn duty to protect federal facilities and those within them."[1]

---

1   Lisa Balick and Jenny Young, "Homeland Security Chief in Portland, Condemns Protests," *KOIN News*, July 16, 2020, koin.com.

Mere hours before these comments, at 2:00 a.m., Mark Pettibone walks back to his car after the night's protest. He is angry for George Floyd, for Breonna Taylor, and for so many others who did not get justice.

A van halts in his path. Four or five men in fatigues hop out. No identifying marks nor any effort to identify their agency. Are they even officials? His mind queues up rumors of abductions and wonders about right-wing militias active in Oregon.

He runs.

They chase him while the van circles around and blocks his path. With no path to escape, he drops to his knees and asks, "Why?"

Nobody answers. Men hurl him into the minivan and speed off. Inside the van they pull a hood over his head They hold his hands and grip his head and neck. He wonders again whether he's been abducted by officers or rogues.

Once the van has stopped, Pettibone glimpses a garage filled with military vehicles. A sense of relief washes over him with confirmation that he's been kidnapped by officials. They photograph him and leave him in a cell for hours. He wonders how many in ICE raids end up like this, but with no end in sight. His captors eventually release him but never identify themselves.

The marshals insist it wasn't them. Pettibone sues to have the records expunged and to find the identity of his captors. DHS admits it agents use unmarked vehicles to pick up protesters. Nobody confirms which secretive police abduct Americans exercising constitutionally protected rights. Many suspect BORTAC.

As of now, Pettibone doesn't know who jumped from an unmarked vehicle that early morning and kidnapped him. "A lingering paranoia and fear have made me hesitant to exercise my rights to the fullest," he writes. "I think that was part of the point of 'the arrest.'"

## Internal Border Precedents

The Great Depression coincided with the Dust Bowl. Drought and poor agricultural practices eroded soil. Desperate families emigrated. It was only a decade after the creation of the US Border Patrol, but the migrants of concern to Western officials were, as Los Angeles police called them, "vagrants and beggars" suffering layers of catastrophes.

The US Supreme Court found states had a right to "exclude from its limits convicts, paupers, idiots and lunatics, and persons likely to become a public charge," a right founded on "the sacred law of defense."[2] States could ban US citizens from neighboring states. Immigration acts in the following years banned (in official terminology) beggars, undesirables, idiots, alcoholics, and anarchists from entering the United States. In 1917 Arizona deputized a couple of thousand citizens to drive 1,300 striking mine workers into New Mexico.[3]

These events provided precedent for the so-called bum blockades California enacted during the Depression. One hundred and thirty-six police were deployed from Los Angeles. They stopped travelers on roads and trains at sixteen entry points along state lines and searched for anyone who looked to have "no visible means of support" (not unlike the old Black Codes forcing African Americans to carry documentation of employment). Officers fingerprinted travelers "who looked like criminals," according to Rose Marie Packard in her 1936 report for *The Nation*. When she asked an officer whether he knew people had a constitutional right to travel, he replied he was acting on orders. The police chief said constitutional rights were of "no benefit to anybody but crooks and criminals."[4]

Los Angeles police terrorization was one part of a two-pronged operation. "Housewives" were encouraged to report "hoboes." In *Storming the Wall*, the migration researcher Todd Miller likened the operation to the "See Something, Say Something" campaign after 9/11. Citizens were engaged in interactive surveillance.

California justified its blockade on grounds that an influx of penniless families from other states meant California did far more than its fair share. Stereotypes painted hundreds of thousands of environmental refugees as lazy and keen to have too many children.[5] When Colorado sent its National Guard to the state line, it was to defend against low-income American neighbors. Florida sent police to its border to establish a "poverty quarantine" to stop those who would "turn to crime for support." American citizens, many of them

2 *Hannibal & St. Joseph Railroad Company v. Husen*, 95 U.S. 465 (1877).

3 Miller, *Storming the Wall*, 140–1.

4 Rasmussen, "LAPD Blocked Dust Bowl Migrants at State Borders," *Los Angeles Times*, March 9, 2003, latimes.com.

5 Miller, *Storming the Wall*, 139.

white, found themselves stamped with some of same derisive labels used for migrants of color who weren't so lucky.

It wasn't the only time American citizens found themselves caught up in border patrols. During World War II, the bracero program balanced a labor shortage by allowing millions of Mexican citizens to work in the United States. Their status was tied to employment. Authorities cracked down after the war and launched Operation Wetback in 1954. Well over a million workers were deported to Mexico. Some were American citizens caught up in the sweeps. Even as Mexican nationals and some US citizens were deported, more were still brought in as braceros.

Today, the H-2 temporary worker program inherits part of the role previously served by the bracero program. The slack is picked up by migrants working for employers who sporadically call DHS to remove leaders or whoever demands timely wages. What these programs and their histories show us is a confused mix of white supremacy, agribusiness, wage suppression, strikebreaking, tenuous employment, and terrorization.

We don't know how the North will respond to the Changes, even if history rhymes. The United States likely won't close interstate borders, but perhaps they will. It isn't hard to imagine new travel restrictions mapping onto race, employment, or documentation status. Perhaps the Schengen Area will fracture as well. The left needs to think about effects of resource scarcity and the energy transition on inequality, the suffering of the working class, xenophobic backlash, and deranged right-wing signals amplified by real or imagined effects of the Changes.

A simple lesson from the War on Terror, where American police departments bought up cheap military surplus: weapons meant for elsewhere will be turned on citizens because they went to a protest or don't own a home. The Global North won't remain unscathed, but it will feel terribly normal.

## Threat Multipliers

The US military is the largest single consumer of hydrocarbons in the world, but much of its emissions are exempt from carbon accounting. Ahead of the Kyoto Protocol, Dick Cheney and a murder of lobbyists pushed an opt-out clause: "Emissions resulting from multilateral operations pursuant to the Charter of the United Nations shall not be included in national totals, but reported separately."[6] In a treaty the United States didn't finally ratify, it wrote itself a blank check for military emissions so long as operations were backed by the Security Council.

Annual US emissions are nearly 6 $GtCO_2$-eq, US military emissions average 70 $MtCO_2$-eq, and that means the military is responsible for a little over 1.2 percent of US emissions.[7] That does not include concrete, steel, bunker fuel used for international transport or multilateral operations, or Scope 3 emissions, all of which the military evades thanks to its Kyoto exemption. A percent might not sound like much, but it's more than most countries. The vast majority of the US military's emissions are from jet fuel, for which there's no clean alternative.[8] It's a terrible offense to the carbon budget from an entity whose primary activity is ensuring access to fossil fuels.

We saw Obama discuss a green navy fleet or sustainable jet fuels, and he's not alone in painting the military green. In her 2020 campaign for the presidency, Senator Elizabeth Warren ran on greening the military. Her plan cited the "Army Net Zero Initiative," a program that doesn't actually aim to eliminate use of fossil fuels.[9] Instead, the army recasts net zero as "a holistic strategy founded upon the Army's long-standing sustainable practices and incorporation of emerging best practices to manage energy, water, and solid waste." Much of the report is about toilets and trash.

Rather than cut budgets and close bases, Warren wanted to "harden the U.S. military against the threat posed by climate

6 "Report of the Conference of Parties on Its Third Session, Held at Kyoto From 1 to 11 December 1997, Addendum, Part Two: Action Taken by the Conference of the Parties at Its Third Session," UNFCCC, March 25, 1998, unfccc.int.

7 Specifically, US emissions totaled 5.79 $GtCO_2$-eq in 2018, and military emissions were 1.267 $GtCO_2$-eq over 2001–18, or on average 70.39 $MtCO_2$-eq. See Ritchie and Roser, "$CO_2$ and Greenhouse Gas Emissions"; and Neta Crawford, "Pentagon Fuel Use, Climate Change, and the Costs of War," Watson Institute, International & Public Affairs, Brown University, November 13, 2019, watson.brown.edu.

8 Crawford, "Pentagon Fuel Use, Climate Change, and the Costs of War," 11.

9 "2015 Progress Report: Army Net Zero Initiative," Assistant Secretary of the Army (Installations, Energy and Environment), August 2016, api.army.mil.

change," and said, "We don't have to choose between a green military and an effective one." She proposed a military climate czar and called for net-zero carbon emissions for all non-combat bases by 2030. No plans were presented for how to accomplish this feat. It's a liberal fantasy of a defensive military linked to a carbon purity wish. But there are no green bombs, nor is there a substitute for jet fuel.

Militaries have studied climate change for a while. In the Cold War, the Pentagon planned for conflict in a thawed Arctic. More recently, in 2004, the UN released a report called "A More Secure World: Our Shared Responsibility." Without using the exact phase "threat multiplier," the report linked extreme ethnic and regional inequalities, civil violence, concentrated young populations, and unemployment. It singled out environmental degradation as an amplifier. "Rarely are environmental concerns factored into security, development or humanitarian strategies," the authors said.[10]

Soon the notion of a threat multiplier caught on. By 2009, the UN treated the fairly young idea as conventional wisdom: "Climate change is often viewed as a 'threat multiplier.'" At President Obama's direction, the 2016 Department of Defense directive 4715.21 established policies "to assess and manage risks associated with the impacts of climate change." The report called for safeguarding the economy, infrastructure, environment, and natural resources. Yes, they put climate change front and center of strategic plans by openly studying how to use the military to protect energy infrastructure. Climate change is a buffet for militaries, border patrols, and arms dealers. Their jargon obfuscates who they consider the threat.

While recounting the generals and contractors he's heard worry about threat multipliers, Todd Miller cut through the handwaving. "More dangerous than climate disruption was the climate migrant. More dangerous than the drought were the people who can't farm because of the drought." After all, if a global food and water crisis can be solved with only a fraction of a US military budget already, alleviating famines with direct aid isn't too difficult. But the climate refugee "was a threat to the very war planes required to enforce the financial and political order where 1 percent of the population wielded more economic power than the rest of the world

---

10   "A More Secure World," United Nations Peacebuilding, December 2, 2004, un.org.

combined."[11] The US military facilitates flows of fuel and capital, which displacement and self-determination threaten.

What the military patrols suppresses overseas, domestic policing mimics at home. Whereas AFRICOM controls and destroys bodies in the Sahel under the pretense of terrorism, DHS controls spaces inside the United States under whatever pretexts it plausibly cobbles together.

## One Hundred Miles

Since its founding after 9/11, DHS oversees border enforcement and houses Customs and Border Patrol along with Immigration and Customs Enforcement. Its budget ballooned to $50 billion, with $16 billion for CBP alone (more than the FBI, US Marshals, and the Bureau of Alcohol, Tobacco, and Firearms combined). CBP patrols the border, while ICE terrorizes migrants who make it past.

One CBP official defined their role as operational control of the border, but clarified: "'Operational control' shall mean the prevention of *all* unlawful entries into the United States, including entries by terrorists, other unlawful aliens, instruments of terrorism, narcotics, and other contraband." Notice the blurring of terrorism, drugs, and migrants.

The vast majority of Americans unwittingly live within an area where border guards hold legally murky authority to conduct searches and set checkpoints. They regularly overextended power. A hundred mile "zone of security" extends from any boundary, either north or south from a border or inland from an ocean. For example, the entire state of Michigan sits within this zone, because it's surrounded by the Great Lakes bordering Canada. Florida, several states in the northeast, and most major cities in California lie within the zone. Within twenty-five miles of a border, CPB agents can enter nonresidential buildings without a warrant.

In the liminal zone, CBP agents regularly ignore court restrictions on vehicle searches without "reasonable suspicion."[12] What constitutes reasonable suspicion is ripe for abuse. When the Obama administration implemented rules against profiling based on race

---

11    Miller, *Storming the Wall*, 67.

12    Chris Rickerd, "Customs and Border Protection's (CBP's) 100-Mile Rule," American Civil Liberties Union, aclu.org.

and ethnicity, immigration enforcement stayed exempt. An official defended the position: "The immigration investigators have said, 'We can't do our job without taking ethnicity into account.'"[13]

Two-thirds of Americans and many undocumented families, about 200 million people, live within that hundred-mile zone. DHS holds extraordinary authority and regularly breaks even the bare constraints placed on its behavior. Agents often extend beyond the hundred miles. At the 2016–17 Dakota Access Pipeline protests at Standing Rock, CBP officers operated well beyond the hundred-mile zone.

In 2020, CBP drones circled a Minnesota town ninety miles from the border, home to the Indigenous Environmental Network, which protests pipelines. Responding to an inquiry from the journalists Yessenia Funes and Dhruv Mehrotra, CBP claimed "it does not patrol pipeline routes."[14] In 2021, a DHS helicopter buzzed protesters who sought to stop the construction of a tar sands pipeline called Line 3, kicking up dust to intimidate hundreds of people.[15] Defense of the homeland means protection of fossil fuel infrastructure.

Portland is eighty miles from the ocean, which returns us to the first story in this chapter. Though details are still withheld as of this writing, Mark Pettibone was likely abducted by CBP. By their own admission DHS and the BORTAC unit were in the area. Unmarked vans and extra-constitutional seizures within a hundred-mile zone fit the DHS operational profile. Repressing protesters exercising constitutional rights is deeply troubling already, but BORTAC's mission is "to respond to terrorist threats of all types anywhere in the world in order to protect our nation's homeland," a hell of a stretch beyond the border.

While border guards transform into secret police, the military turns into border guards. In July 2021 Governor Kristi Noem announced she'd deploy fifty soldiers from the South Dakota

13  Matt Apuzzo and Michael Schmidt, "U.S. to Continue Racial, Ethnic Profiling in Border Policy," *New York Times*, December 5, 2014, nytimes.com.

14  Yessenia Funes and Dhruv Mehrotra, "CBP Drones Flew Near Indigenous Pipeline Activists' Homes," Gizmodo, September 18, 2020, gizmodo.com.

15  Alleen Brown and Sam Richards, "Low-Flying DHS Helicopter Showers Line 3 Protests with Debris," *The Intercept,* June 8, 2021, the intercept.com.

National Guard to Texas, where they would guard the United States–Mexico border. In an incredible twist, a Tennessee billionaire, the Republican megadonor Willis Johnson, paid for the operation. At the same time, the South Dakota–based land back attorney Bruce Ellison exposed further plans to use the South Dakota National Guard against water protectors resisting pipelines like Line 3 in Minnesota. According to Ellison, under certain conditions the troops would be authorized to use lethal force.[16] Don't lose sight of what happened here: soldiers targeted citizens over oil.

Imagine checkpoints harassing people in Phoenix, barely more than a hundred miles from the border but suddenly justified were the city to run short on water. Think of the terror erected at every artery in or out of a city under the guise of "public safety." Keep imagining the scenarios within or near the hundred-mile zone in which two-thirds of Americans live. If a city were flooded, CBP and mercenaries would roll in. If American states have closed borders to other states before, what happens if impoverished or nonwhite southerners start displacing?[17] Will liberals in the Northeast or Midwest feel red state residents deserve to suffer on account of regressive politics that led to the Changes? What if DHS "secures" Indigenous reservations to save pipelines? How will BORTAC respond when protesters appear anywhere remotely close to sea-walls erected to protect Washington, DC, or Wall Street? After Covid-19, we should consider how new variants or other viruses amplify economic stress and feed xenophobias. What new sources of barbarism shall we invent, now to hurt the other, later to turn against ourselves?

## American Climate Migrants

Fifty thousand Californians were displaced by the Camp Fire, which devastated a town called Paradise on November 8, 2018. In a matter of hours, too quick for many to hear an evacuation order, at least eighty-five people died. Nearly nineteen thousand buildings were destroyed. So far, the town has recovered only around a quarter of its population, but whether out of resilience

---

16 "Lethal Force Against Pipeline Protests? Documents Reveal Shocking South Dakota Plans for National Guard," *Democracy Now!*, July 2, 2021, democracynow.org.

17 Several examples inspired by Miller, *Storming the Wall*.

or foolishness, three years later headlines declared Paradise the fastest growing city in California.

Residents were displaced to the neighboring city of Chico. They leased apartments if possible. Rising demand and housing prices forced others to sleep in trailers or tents. Half of senior citizens left the region. The bulk of survivors remained in central California, but at least some of them ended up in every state throughout the United States. Some survivors built a new life along the Great Lakes or the Eastern Seaboard, never again to fear the fires haunting those who returned. Do they think of themselves as climate migrants?

In her book on climate curricula taught to US schoolchildren, the reporter Katie Worth recounted the answer given by a child who barely escaped the Camp Fire. His school relocated from Paradise to a hardware store in Chico. Lunch was served at the checkout counter and classes held in the aisle. Amid shelves of ceiling fans and lights a teacher asked students to describe how climate change impacted them. This child, this refugee five months removed from a harrowing flight, answered: "It hasn't affected my life at all yet . . . I don't know if it will do anything to my life in fifty years because I don't know if I believe it yet."[18]

This child mirrored the answer adults around him would give. What stood out was "I don't know if I believe it yet," as if belief influences whether he will be affected. Normal for a kid. Dangerous for adults.

By temperature and precipitation, the human climate niche currently sits in the American South and lower Midwest, stretching from the Plains to the East Coast. Over the next fifty years, the human climate niche will shift north. By 2070, the suitable zone will be thoroughly Midwestern. In the worst-case scenarios, the suitable zone would stretch up to the top of Michigan and into Canada, areas so cold right now the Great Lakes freeze over close to shore in winter months.

By midcentury, Missouri will feel as warm as Louisiana today, per data from the Rhodium Group in collaboration with *ProPublica* and the *New York Times Magazine*. Wildfires will scorch Western states, but much of Florida, Georgia, South Carolina, and even Minnesota will see wildfires as well. Property will go underwater. Crop yields in the southern Black Belt will decline anywhere from

---

18 As quoted in Katie Worth, *Miseducation* (New York, NY: Columbia Global Reports, 2021), 150–1.

a few percentage points to as much as half, while, in the Midwest, yields will improve but not enough to balance. Even in moderate emissions scenarios, the South faces severe economic losses in 2040–60. Cascading losses might not alter preparations. After all, when a North Carolina commission reported sea level could rise by a meter, meaning the report threatened property values, the state simply banned policymaking based on forecasts.

People will move north if afforded the luxury. The Northeast and Midwest will grow. Perhaps we'll prefer living near freshwater in the Great Lakes. Wealthy Californians will think it better to move a mile high to Denver or to upstate New York where there are fewer fires. What of those who can't move?

During the pandemic, in my exasperation at anti-vaxxers it was tempting to throw my hands up and say of the reactionaries, "Let them die." I wonder whether, during the Changes, we'll see affluent Americans who move, along with those lucky enough to be born in northern latitudes, viewing the hot, flooding South or scorched West as suffering for their own faults. Like everything else in American life, the ability to move will follow racial contours. Inhospitable areas with plummeting property values will grow more undocumented, Black, and brown. Vestiges of intergenerational wealth from home equity will disappear and trap residents. Police will enjoy themselves. Mercenaries will cash in. People of color and workers will suffer most. In recent memory, we've seen omens of future repression.

## Lessons in Repression: Katrina

The hurricane made landfall in Louisiana on August 29, 2005, and breached levees around New Orleans. Hurricane Katrina reached Category 5 status over the Gulf of Mexico, but by the time the levees failed and flooded the city, the storm slowed. A fifth of the city was underwater, including much of the predominately Black Lower Ninth Ward. The official death toll shot to 1,833, not counting hundreds missing.

The Bush administration responded slowly, at great cost to life, while racist rumors spread about the condition of thousands who sheltered in the Superdome sports stadium. Myths included murders and rapes, infants with throats slit, gangs in charge, and so much shit smeared on the ground that corpses were tossed into the excrement. The fiction spread via television news. Reports of looters and inhuman filth fantasies surely justified violence from mercenaries and militias on soaked ground.

Within a couple weeks, 235 private security firms registered operations in Louisiana, per a report by Jeremy Scahill.[19] The most notorious was Blackwater, under contract with DHS. Some mercenaries worked government contracts while others guarded private property for the wealthy. The Bush administration sought unsuccessfully to repeal the Posse Comitatus Act to use soldiers in American cities; thus private contractors, often ex-soldiers, provided a workaround. Troops recently back from the criminal invasion of Iraq could make far more money with Blackwater brutalizing the people of New Orleans.

A mercenary contracted by a rich hotelier told of "Black gang-bangers" firing on his position. "I dropped the phone and returned fire . . . After that, all I heard was moaning and screaming, and the shooting stopped. That was it. Enough said." Another mercenary brought in to guard a luxury area of town tapped his M-16 as he explained, "Most Americans, when they see these things, that's enough to scare them."

Mercenaries traveled in unmarked cars without license plates while others traveled in SUVs with tinted windows. Some took over bars or homes and set up bases. One heard his assignment and asked, "When they told me New Orleans, I said, 'What country is that in?'" Scahill asked Blackwater mercenaries bearing assault weapons and ammo strapped to flak jackets what their mission was. One answered that it was "securing neighborhoods" and "confronting criminals." Another said his fellow mercenary "was even deputized by the governor of the state of Louisiana. We can make arrests and use lethal force if we deem it necessary."

Louisiana governor Kathleen Blanco proudly announced the several hundred National Guardsmen deployed to New Orleans after Katrina "know how to shoot and kill" the "hoodlums."[20] On September 6, a group of police officers opened fire on a family on the Danziger Bridge. The officers wounded four and killed two, a minor and a mentally disabled adult.

---

19   Jeremy Scahill, "Blackwater Down," *The Nation*, September 21, 2005, thenation.com.

20   AFP/Reuters, "Troops Told 'Shoot to Kill' in New Orleans," *ABC News*, September 2, 2005, abc.net.au.

White militias patrolled neighborhoods. "We shoot looters," announced one sign. Reginald Bell, who is Black, recalled white men aiming guns at him from a balcony near his home who shouted, "We don't want your kind around here!" The next day, the same group threatened his life again at gunpoint. Without electricity or protection from vigilantes, Bell said, "We were like sitting ducks. I slept with a butcher knife and a hatchet under my pillow."[21]

Donnell Herrington was ambushed along with his cousin and a friend three days after the hurricane. The three Black men were in Algiers Point, a wealthy and largely white area. Without warning, three white men opened fire with shotguns, hitting Herrington in the jugular vein and then again in the back, all for the offense of walking on dry land toward an evacuation terminal. As Herrington stumbled in one direction, the other two, ages seventeen and eighteen, fled from the men yelling, "We got you, [racist slur]! We got you, [racist slur]!"

The seventeen-year-old later recalled, "They said they was gonna tie us up, put us in the back of the truck and burn us. They was gonna make us suffer . . . I thought I was gonna die." Herrington, wounded and struggling to survive, heard the assailants shout, "Get him! Get that [racist slur]!"

Algiers Point stayed dry. The Coast Guard set up a makeshift evacuation terminal, so Black residents evacuated through the predominantly white enclave. A few dozen white residents gathered shotguns, Uzis, pistols, and rifles and formed an impromptu militia.

The journalist A. C. Thompson investigated the killings that followed and pointed to a carjacking origin story for one militia.[22] "The kid whacked me," recounted the vigilante leader. "Hit me on the side of the head." This was the motive for what followed. "I'm not a racist," he insisted. "I'm a classist. I want to live around people who want the same things as me." He told Thompson about imagining "goons" would kill his mother and how he screamed at a Black man, "I don't want you passing by my house!" It was this vigilante who told Reginald Bell, "Well, we don't want you around here. You loot, we shoot."

---

21  Trymaine Lee, "Tales of Post-Katrina Violence Go From Rumor to Fact," *New York Times*, August 26, 2010, nytimes.com.

22  A. C. Thompson, "Katrina's Hidden Race War," *The Nation*, December 17, 2008, thenation.com.

A second militiaman, who admitted on camera to shooting someone in this timeframe, felt angry the National Guard used Algiers Point for evacuations. "Hoodlums from the Lower Ninth Ward and that part of the city" were massing in his neighborhood. "I'm not a prejudiced individual, but you just know the outlaws who are up to no good," he reasoned. "You can see it in their eyes."

On camera, a third vigilante treated the episode like sport. "It was great! It was like pheasant season in South Dakota. If it moved, you shot it." He sent "looters" away "full of buckshot," when his machismo erupted uncensored. "You know what? Algiers Point is not a pussy community."

One woman told Thompson that her uncle was one of the vigilantes. "My uncle was very excited that it was a free-for-all—white against Black—that he could participate in," she explained. "For him, the opportunity to hunt Black people was a joy . . . They didn't want any of the 'ghetto [racist slur]' coming over."

Nobody knows how many were killed or maimed by vigilantes, police, soldiers, and mercenaries in this time. Authorities never investigated most of the suspicious deaths. One white resident said the vigilantes had little to fear from shooting African American survivors: "No jury would convict." Indeed, police would not even hassle.

"As you look back on it," recalled an army lieutenant general leading military operations after the hurricane, "it looked like the city was under siege."[23] One survivor, Malik Rahim, asked, "How can you remove the scars from the eyes of all the children who witnessed these atrocities?"

In the end, it was forgotten, and in that way, it was prophecy. After Katrina there came rebuilding and soaring property values. Many who survived the storm and terror were priced out through gentrification. Such violence erupted from vigilantes, soldiers, police, and mercenaries in a major metropolitan American city in full view of authorities with nearly nonstop media coverage.

One Blackwater mercenary in New Orleans delivered a prophecy for future environmental disasters. "This is a trend. You're going to see a lot more guys like us in these situations."[24]

---

23   Lee, "Tales of Post-Katrina Violence."
24   Scahill, "Blackwater Down."

## Lessons in Repression: Pipelines and Mercenaries

Whereas DHS's remit is terrorism and borders, mercenaries provide an additional layer of deniability. Before the Keystone XL pipeline's cancellation (for now) in 2021, law enforcement viewed Indigenous-led protesters as a "domestic terrorism" threat to be stopped "by any means." The pipeline would have transported Canadian tar sands oil through Montana, South Dakota, and Nebraska. Protests against the project began as early as 2009. A decade later, documents exposed a relationship between TC Energy (formerly TransCanada) and authorities.

One law enforcement briefing aimed to "initially deny access to the property by protestors and keep them as far away [from] the contested locations as possible by any means." The Bureau of Land Management sent ten armed officers to coordinate with local police and deny protesters access to federal land. The US attorney's office prepared a "critical incident response team" to fight "domestic terrorism or threats to critical infrastructure."[25]

Before Keystone XL, Indigenous-led protests against the Dakota Access Pipeline at Standing Rock faced down the police's water cannons, less-lethal rounds, and tear gas, as well as mercenaries. The private security firm TigerSwan collaborated with police against DAPL protesters, which internal communications compared to "jihadist insurgency" and described as "an ideologically driven insurgency with a strong religious component."

According to documents obtained by *The Intercept*, TigerSwan thought DAPL protesters would "follow a post-insurgency model" after the protest ended. Since the same protesters reappeared at Keystone XL (activists often join subsequent protests) they are repeat offenders treated as insurgents by mercenaries returning from Iraq. TigerSwan communications prepared intelligence on "the battlefield" to defeat "pipeline insurgencies."

Mercenaries took photographs and license plate numbers. They infiltrated camps in multiple states using false names and developed relationships with protesters. Like other mercenary firms, TigerSwan was founded by a retired army commander during the Iraq War who aimed to compete with Eric Prince's Blackwater (now

---

25  Sam Levin and Will Parrish, "Keystone XL: Police Discussed Stopping Anti-Pipeline Activists 'By Any Means,'" *The Guardian*, accessed May 6, 2022, theguardian.com.

Academi). TigerSwan has offices throughout the Middle East and Latin America. While surveilling protesters at Standing Rock, the company shared intelligence with, among others, the FBI, the Joint Terrorism Task Force, and DHS.

*The Intercept* noted TigerSwan's approach to "find, fix, and eliminate" pipeline threats mimics military special forces language "find, fix, finish," which refers to killing. Its communications are rife with military jargon. Protesters "stockpile" signs. Camps are a "battlefield" full of insurgents. A protester is a "person of interest." Nonviolent direct actions are "attacks." A rift among members is an "operational weakness." An Earth First magazine provides "TTP's [tactics, techniques, and procedures] for violent activity." When a Palestinian American was discovered in the camp, TigerSwan reported, "The movement's involvement with Islamic individuals is a dynamic that requires further examination."[26] Dehumanizing protesters as foreign agents is a pretext for assault.

## Lessons in Repression: Casting Environmentalists as Threats

Several hundred environmental activists are murdered each year. Global Witness recorded 227 killed in 2020, a record high for a second consecutive year. Colombia, Mexico, and the Philippines accounted for half. The logging sector tallied the most murders, followed by dams, mining, agribusiness, and programs transitioning from illegal to legal crops. A vast majority of perpetrators were hired guns, followed by militias, militaries, and police. Global Witness found few cases in which crimes were brought to court, and when they are it's usually trigger-men.[27] Occasionally, there's a glimmer of justice.

Berta Isabel Cáceres was a community leader from the Indigenous Lenca people in Honduras. She led a campaign against the building of a dam on the Gualcarque River, which is spiritually significant for the Lenca people. She filed legal complaints against

26  Alleen Brown, Will Parrish, and Alice Speri, "Leaked Documents Reveal Counterterrorism Tactics Used at Standing Rock to 'Defeat Pipeline Insurgencies,'" *The Intercept*, May 27, 2017, theintercept.com.

27  "Last Line of Defense," Global Witness, September 13, 2021, globalwitness.org.

the Agua Zarca Dam and obstructed roads in the area. One Chinese contractor involved in construction withdrew, while a Honduran company remained determined. When Cáceres was recognized with the Goldman Environmental Prize in 2015, the journalist Katie Livingstone noted the "grim foretelling of her own fate" when Cáceres "dedicated the award to 'the martyrs' who had given their lives to protect Honduras's rivers, lands and mineral resources."

Gunmen killed Cáceres in her home on March 3, 2016. The murder shocked the nation, and the president called it a "crime against Honduras." Finally, in 2019, seven men were sentenced for her murder. In July 2021, the vice president of the company was convicted of masterminding the murder, providing logistical support, and paying the hitmen. It was a rare moment of justice in a system in which every incentive is set against environmentalists.

More recently, attorney Steven Donziger faced a strange private prosecution conducted by a law firm with ties to Chevron after he secured a $9.5 billion judgment against the company. Donziger represented thirty thousand Ecuadoran farmers and Indigenous people poisoned by Texaco's pollution. Chevron inherited liability when it acquired Texaco. In a bizarre case of corporate prosecution due to a loophole allowing a judge to appoint a private prosecutor when the government declined to prosecute, Donziger faced intimidation, three years of house arrest, and even jail time as payback for seeking justice.[28] The case also demonstrated the clarity with which the fossil fuel industry understands their liability. After years of injustice, during the final days of writing this book, Donziger was freed.

Wins are rare. The right busies itself with bogeymen. It frets over antifascists or "cultural Marxists" (among the more openly anti-Semitic terms) as a pretext for repression. This is why Adorno said fascist propaganda attacks bogies rather than real opponents. That's what happened in the introductory story of imaginary anti-fascists setting fires near Portland. When arguing in good faith, there's a legible relationship between premises and inferences. Not so in the right's oratorical exhibitions. Instead, we find ideas linked by the barest similarities, per Adorno, "often through association by employing the same characteristic word in two propositions

---

28 Walker Bragman and David Sirota, "The Government Gave Big Oil the Power to Prosecute Its Biggest Critic," *American Prospect,* July 14, 2020, prospect.org.

which are logically quite unrelated."[29] Soon enough, the credulous audience can't tell the difference between socialism and national socialism.

The accusations are reaction formations, what Melanie Klein called splitting and projective identification. When something is perceived as both bad and good, the ego splits the object into good and bad parts. The ego splits as well and identifies with the idealized good while it exaggerates and condemns the bad. Splitting and projective identification underlie the fear.[30]

The right frets over bogies. It's legible when read as projection. The enemy (Muslim, Jew, person of color, environmentalist, or socialist) *must* be vile to deserve punishment, disenfranchisement, expulsion, or torment, else the right's own violent ideation could feel alarming. Once the right sets the terms, liberals or depoliticized journalists and bureaucrats adopt the framing. The following examples exhibit a range of persecutory fantasies underwriting sadistic enjoyment.

In 2005 the Australian government published a booklet titled *Preventing Violent Extremism and Radicalisation in Australia.* In a section on violent extremism, "environmental concerns" are grouped with religious ideologies and ethnic or separatist causes. Another section warns of violence related to "animal liberation, environmental activism or anti-gun-control." The document develops an absurd, cartoonish "case study" of an imaginary person named Karen. "Karen grew up in a loving family who never participated in activism of any sort." But she went to university and got involved in the "alternative music scene, student politics and left-wing activism." Karen dropped out of university to devote herself to disruption of the logging industry. She was repeatedly arrested for property damage and assaulting police. She eventually returned to university and works for a mainstream environmental organization, and, we are told, she now believes "illegal or aggressive direct-action campaigns only produce short-term solutions."[31]

In November 2019, a pamphlet for teachers, government officials, and police calls the nonviolent group Extinction Rebellion (XR)

29  Adorno, *The Stars Down to Earth*, 165.

30  See Melanie Klein, "Notes on Some Schizoid Mechanisms," in *The Writings of Melanie Klein*, vol. III, ed. Roger Money-Kyrle, (New York: The Free Press, 1984).

31  "Preventing Violent Extremism and Radicalisation in Australia," Australian Government Attorney-General's Department, 2015.

an extremist organization alongside neo-Nazis.[32] Teachers should watch for students who engage in planned school walkouts, a reference to Greta Thunberg's Fridays for Future strikes. Look out for students "writing environmentally themed graffiti," says the resource produced by counterterrorism police. What threat did they pose? The pamphlet explained that, while XR is non-violent, it "encourages other law-breaking activities" and is "an anti-establishment philosophy that seeks system change."

In July 2021, the American Psychological Association published an article grouping environmentalists, socialists, and communists together as extremists and possible domestic terrorists to be studied.[33] "Some members of the anti-fascist political protest movement Antifa have enacted violent extremist crimes," the article asserted. The underlying study to which it pointed described "left-wing attacks" as property vandalism, not real violence against humans, but the APA author left this out. The study listed as acts of terrorism: "pursue environmental or animal rights issues," "espouse pro-communist or pro-socialist beliefs," and "support a decentralized social and political system."[34] But when discussing QAnon conspiracy believers arrested for ideologically motivated crimes, the APA cited a study characterizing them as mentally ill or traumatized. In other words, the APA blended the right's violence, which it greets with understanding, with socialists who it calls terrorists.

In April 2021, Ohio escalated legal protections for "critical infrastructure." Almost all nineteen categories of critical infrastructure protected involved fossil fuel extraction, refining, or transport, and petroleum refineries were at the top of the protected list.[35] Revisions to the code included "an owner or operator of a critical infrastructure facility may elect to commence a civil action . . . against any person who willfully causes damage to the critical infrastructure

---

32  Vikram Dodd and Jamie Grierson, "Terrorism Police List Extinction Rebellion as Extremist Ideology," *The Guardian*, January 10, 2022, theguardian.com.

33  Zara Abrams, "Deradicalizing Domestic Extremists," *Monitor on Psychology*, July 1, 2021, apa.org.

34  Seth Jones and Catrina Doxsee, "The Escalating Terrorism Problem in the United States," Center for Strategic and International Studies, June 17, 2020, csis.org.

35  "Title 29, Chapter 2911, Section 2911.21," Ohio Revised Code, April 12, 2021, codes.ohio.gov.

facility." Environmental organizations or non-present activists could be held "vicariously liable" if they "directed, authorized, facilitated, or encouraged the person to cause damage to the critical infrastructure facility."[36] "Improperly tamper" was redefined as "to change physical location or the physical condition of the property," and trespassing was made a felony.

The bill was Ohio's version of a 2017 American Legislative Exchange Council (ALEC) shell text. ALEC creates draft bills for conservatives to modify and implement in their states, and it has taken a lot of fossil fuel money. Ohio is the fourteenth state to pass some version of this anti-protest bill.[37]

Montana was the fifteenth state to ban types of climate protest when in the spring of 2021 it targeted persons who "impede or inhibit operations." ACLU attorney Vera Eidelman said the situation grows worse: "What we are seeing this year with the huge wave of anti-protest bills is that it is consistent with, but frankly worse than, what we saw for the previous four years under Trump." Some ninety anti-protest bills were introduced across thirty-six states in one year.[38]

Arkansas's version, like Ohio's, threatened protesters with a $10,000 fine. Montana's threatened a $150,000 fine and a thirty-year prison sentence. If an organization was found guilty of encouraging activism, it could be fined $1.5 million.[39]

It wouldn't be a stretch to argue that this book you hold, according to these deranged laws, constitutes encouragement to damage infrastructure. The lines of legal and illegal speech are vague, which is an intended chilling point. Just as conservative legislatures rushed bills to protect reactionaries who rammed vehicles into protesters during the George Floyd protests against police murder, they elevated protest, speech, and environmental activism to the status of felonies at the precise moment the public awakened.

---

36    "Substitute Senate Bill Number 33," General Assembly of the State of Ohio, 2021, legiscan.com.

37    Alexander Kaufman, "Ohio Quietly Passes a Bill That Could Bankrupt Churches Linked to Fossil Fuel Protest," *HuffPost*, December 19, 2020, huffpost.com.

38    Jeremy Miller, "Republicans Want to Make Protesting a Crime," Sierra Club, May 25, 2021, sierraclub.org.

39    Ibid.

## Lessons in Repression: Eco-Fascist Fantasies

"My whole life I have been preparing for a future that currently doesn't exist," wrote a twenty-one-year-old college student. "The environment is getting worse by the year. If you take nothing else from this document, remember this: INACTION IS A CHOICE."

He grounded what he was about to do in overlapping crises. "Our lifestyle is destroying the environment of our country. The decimation of the environment is creating a massive burden for future generations." He wrote in his manifesto about a classic Dr. Seuss character, the Lorax, who speaks for the trees, as he fumed at corporations that over-harvested the land, poisoned watersheds, and leveled forests. Americans refused to stop. "God damn most of y'all are just too stubborn to change your lifestyle," he whined. "So the next logical step is to decrease the number of people in America using resources. If we can get rid of enough people, then our way of life can become more sustainable."

It was the morning of August 3, 2019. He posted his manifesto online and gathered weapons. In the document titled "The Inconvenient Truth," after Al Gore's climate documentary, he complained about a "Hispanic invasion of Texas," a loss of jobs, and a "Great Replacement" of whites. He said Democrats would enact a "political coup" via demographic change. He said he was "simply defending my country from cultural and ethnic replacement."

A fusion of fears swirled in his mind as he entered a Walmart in the largely Latino community of El Paso, Texas, that morning. At 10:40 a.m., he raised his gun. It would overheat quickly, he explained in his manifesto, so he'd wear heat-resistant gloves to protect himself as he killed and killed and killed.

Twenty-three victims died, one a child. Another couple of dozen were hit, including several children, one just a toddler.

"Many people that think that the fight for America is already lost," he concluded in his manifesto. "They couldn't be more wrong. This is just the beginning of the fight for America and Europe. I am honored to head the fight to reclaim my country from destruction."

The El Paso shooter didn't pioneer new terrain by linking environmentalism to ethnic cleansing. Environmentalism is now so thoroughly associated with liberal or left politics that it's hard to think of the Sierra Club as once a bastion of nationalism, but the history suggests overlaps in right and green fantasies. For a time, nationalists laundered immigration restriction in the language of ecological protection.

The Southern Poverty Law Center described John Tanton, a Sierra Club and Planned Parenthood chapters founder, as "the racist architect of the modern anti-immigrant movement." The *New York Times* reported that his organization Federation for American Immigration Reform "hoped to enlist unions concerned about wage erosion, environmentalists concerned about pollution and sprawl, and Blacks concerned about competition for housing, jobs and schools."[40] Over the decades, he bridged a type of bastardized intersectionality and eco-consciousness with reactionary nativism.

In a 1993 letter to ecologist and fellow ethno-nationalist Garrett Hardin, Tanton wrote, "I've come to the point of view that for European-American society and culture to persist requires a European-American majority, and a clear one at that." Hardin was most famous for his "tragedy of the commons," the idea being that if resources are available in common to everyone in a community, morally defective takers have an incentive to consume more than their fair share. In his 1974 essay "Lifeboat Ethics: The Case Against Helping the Poor," Hardin swapped "spaceship Earth," the favored metaphor of environmentalists, for a lifeboat. Suppose you sit in a lifeboat that has space for a few more while a much larger group desperately wades in the water. Allowing in a few more would save those lives, but who are we to choose? We might be tempted toward Christian ideals of being "our brother's keeper" or a Marxist idea of "to each according to his need," but Hardin cautioned that acting on such ethics would drown us all. Should we allow additional passengers on a "first come, first serve" basis? Would we have the right to defend ourselves against any who wished to climb aboard? Some pathetic soul might even wish to give up their spot aboard. "My reply is simple: 'Get out and yield your place to others,'" Hardin wrote tersely. "This may solve the problem of the guilt-ridden person's conscience, but it does not change the ethics of the lifeboat."

The takers Hardin feared were not the top 1 percent, who emit more than the bottom half, or the ten richest men, who own more

---

40   Jason DeParle, "The Anti-Immigration Crusader," *New York Times*, April 17, 2011, nytimes.com.

wealth than the poorest three billion people.[41] He wrote with an eye toward people of color migrating.

In the nineties, Tanton wrote that, while immigrants could assimilate to American society, too many would ruin America by making it mirror Latin American states. In another letter, echoing Hardin's fear of the poor out-reproducing wealthy whites, Tanton asked, "Do we leave it to individuals to decide that they are the intelligent ones who should have more kids? And more troublesome, what about the less intelligent, who logically should have less?"[42]

By 1998, Tanton's agenda was narrowly defeated within the Sierra Club. The organization barely faced down another nativist takeover of its board in the early aughts, and to their credit they've acknowledged that troubling history. In fact, much of the story described here is published on the organization's website.[43] The lesson is this: conservatives aren't bound to reactionary climate denial and easily couch racism in environmentalism.

Tanton died not three weeks before the El Paso shooting. Had he lived, he might have felt pleased with the ethos expressed by the shooter. Tanton fretted over culture "transplanted from Latin America to California," whereas the El Paso shooter worried about "the Hispanic invasion of Texas." The former spoke of the "uncomfortable truth" of mass migration. The latter called it "The Inconvenient Truth."

Only months before the massacre in El Paso, another monster shot up two mosques in Christchurch, New Zealand, murdering fifty-one people and wounding forty. He livestreamed the attack on Facebook. According to the El Paso shooter, the Christchurch shooter's seventy-four-page manifesto was the inspiration for his concern about the Great Replacement. "This is ethnic replacement," said the Christchurch shooter. "This is cultural replacement. This is racial replacement. This is WHITE GENOCIDE."

---

41   Isak Stoddard et al., "Three Decades of Climate Mitigation," *Annual Review of Environment and Resources* 46 (2021): 653–89, 656; and "Ten Richest Men Double Their Fortunes in Pandemic While Incomes of 99 Percent of Humanity Fall," Oxfam International, January 17, 2022, oxfam.org.

42   "John Tanton," Southern Poverty Law Center, accessed May 6, 2022, splcenter.org.

43   Hop Hopkins, "How the Sierra Club's History with Immigrant Rights Is Shaping Our Future," Sierra Club, November 2, 2018, sierraclub .org.

He was the primary inspiration when an eighteen-year-old live-streamed his mass shooting at a Buffalo, New York, grocery store in May 2022 that cost ten their lives. In his manifesto calling himself an eco-fascist, the Buffalo shooter obsessed over replacement rates for whites and said he selected his target due to the area's high Black population. "Green nationalism is the only true nationalism," he wrote in a section about the need to control migration and population size. "For too long we have allowed the left to co-opt the environmentalist movement to serve their own needs."

The Christchurch shooter, in turn, claimed inspiration from a white supremacist who murdered nine people, all African Americans, at a church in Charleston, South Carolina, in 2015, as well as one who blew up a van killing eight in Oslo before he murdered sixty-nine young Labour Party activists at a camp in 2011.

The idea of a Great Replacement was also echoed in the protest in Charlottesville, North Carolina, in 2017. Men carrying tiki torches and Nazi flags chanted, "You will not replace us!" which mutated to "Jews will not replace us." The concept is often attributed to the contemporary French writer Renaud Camus, but it stretches back further than the 2011 article "Le Grand Remplacement." I distinctly remember hearing the same fears in school and church as a child, though nobody called it by that name. "The great replacement is very simple," Camus once said. "You have one people, and in the space of a generation you have a different people."

Proponents of the Great Replacement fret over reproduction rates of races and faiths. Thanks in part to the book *Eurabia: The Euro-Arab Axis* by Bat Ye'or (whose real name is Gisèle Littman), there exists a conspiracy theory positing a Muslim takeover.[44] In *Eurabia*, white Christians only appear in charge but are actually subservient. The Eurabia theory claims parts of Europe are under sharia law, which some American reactionaries apparently believe about certain US cities with large African migrant communities.

Eco-fascists today descend from decades of nativist environmentalism, yet we tend to think of environmentalism as a liberal or left phenomenon. Why? Richard Seymour in *The Disenchanted Earth* noticed two significant reasons. First, since the seventies, the environmental movement turned its gaze to global warming and away from local conservation of land. "In one form of ecofascist idiom, what one conserves is the habitat of the ethos," Seymour explained.

---

44  Malm et al., *White Skin, Black Fuel*, 47–51.

"The land, from that point of view, is where nature and culture meet in a form of life that must be protected from 'invader' species and people." Second, the left winning control of the movement pushed aside the right and allowed resentment to fester out of sight. When the Earth First! founder Dave Foreman argued migration ought to be restricted, welfare denied to people with too many children, and foreign aid to Africa cut, leading to him being pushed out, he seethed about "infiltration" of the movement by the left.[45] Is that so different from the Buffalo shooter saying the right must take back the environmental movement co-opted by the left?

In *Overheated*, Kate Aronoff pointed to the Christchurch and El Paso shooters as an eco-fascism shifting its ideological footing. "That stark partisan divide probably won't last," she warned on polarities of nationalism and environmentalism. "Environmental concerns among young conservatives are growing nearly as fast as they are among young people in general."[46] The conservative pollster Frank Luntz recently found more than half of younger GOP voters are getting worried about climate. This is why rising belief in climate change does not solve the problem of denial, especially in its violent forms. Ironically, Luntz is the one who promoted "climate change" as a less frightening term (unlikely to trigger regulation) than "global warming" during the George W. Bush administration. Now everyone calls it climate change, but where will that lead young American conservatives? If Americans follow the trend Malm identified among European right-wingers, we'll see immigration discussed every time there's a comment on climate.

The pandemic showed how public health or safety mainstreams xenophobia. Early in 2019, the Trump administration implemented a "Remain in Mexico" policy to illegally deny asylum. As Covid-19 lockdowns began, the administration ramped up repression with Title 42, which lets the executive deny entry to asylum seekers if a threat of disease exists. The administration used Title 42 to expel people from Haiti, Mexico, and the Northern Triangle.

Upon taking office, Biden suspended the use of Title 42. By summer, the courts ordered the administration to restart enforcement. The Biden administration press secretary Jen Psaki said it was unfortunate, "but we also believe in following the

45   Richard Seymour, *The Disenchanted Earth* (London: Indigo Press, 2022), 121–2.

46   Aronoff, *Overheated*, 157.

law."[47] Well over a million people were expelled under Title 42 in the first year and a half of the pandemic.[48] The episode reminds us we're in the dark as to how the judiciary will reconfigure human rights in the Changes. More recently, Texas Governor Greg Abbott named *Plyler v. Doe*, a 1982 Supreme Court case requiring states to fund education for undocumented children, as one he wishes to revise. Imagine conservative grassroots pressure mounting against equal protection as the climate migrants pick up. Would liberals check those human rights abuses? Just as the Biden administration continued the Trump administration's unjustified expulsions and asylum denials under Title 42, so, too, has Biden continued his predecessor's use of migrant concentration camps. What is a conservative Supreme Court's decision but the laundering of a manifesto's rambling in erudite polish?

The pandemic was pretext for brutalization in Europe as well. Maltese officials said Covid-19 prevented them from taking any more migrants. So when a boat crossing the Mediterranean sent a distress signal, officials left the stranded passengers floating for five days. Several died of dehydration. At the same time, officials paid private operators to move the boat to Libya. From Italy to Bangladesh to Malaysia, the virus was an excuse to close ports and strand people in the water.

Where the conservative gleefully enjoys sadism, liberals turn a blind eye to humanitarian disaster during liberal administrations. Harsha Walia pointed to the temporary language we use for migration ("migrant invasion" or "migrant crisis"), writing, "American liberals may demand an end to excessive violence against Latinx migrants and refugees, exemplified in their opposition to concentration camps or family separation, but they rarely locate immigration and border policies within broader systemic forces."[49] In short, the same Not in My Backyard selfishness driving socially conscious liberals to oppose low-income housing in their neighborhoods may turn out, in the Changes, to call for both solar panels and smart borders.

---

47   "Rights Groups' Warning as Trump's Remain in Mexico Policy Restored," *BBC News*, March 2021, bbc.com.

48   "A Guide to Title 42 Expulsions at the Border," American Immigration Council, August 2021, americanimmigrationcouncil.org.

49   Walia, *Border & Rule*, 3.

## This Storm Is What We Call Progress

Rampant wealth inequality today foreshadows disaffected and downwardly mobile populations turning anger against migrants. Capitalist logic leans on race for bureaucratically routinizing quotidian violence and lopsided distributions of resources and labor. The Changes won't fundamentally alter our drives but will, at all times, exacerbate prior tendencies. It's a bleak but not inevitable path.

Marx said that it's not our consciousness that creates our social existence, but rather our social existence that creates consciousness. The left must urgently disable material roots for scarcity thinking to which fascism appeals. We need organized workplaces with unions. In the United States, where healthcare is often tied to an employer such that people are trapped in jobs for the sake of family medical needs, we must demand universal health care so that people can move in the event that fires encroach or the seas rise. We'll need to force reactionary nativists, against their will if necessary, to have housing security, enriching work, living wages, and dignity.

We are around a third to halfway through this century's warming, yet we still think of now as normal and the Changes as far off in the future. But to paraphrase Gibson, the barbarism is already here. It's just not evenly distributed.

When the IPCC released its 2018 special report on the vast difference between 1.5°C and an unthinkable 2°C, journalists quickly settled on an appropriate metaphor: we would need a World War II–type of mobilization to cut emissions in half in twelve years. Not much time, the nihilists jeered at the urgency. In the 2030s, reactionaries will say they remember when all the scientists said the world would end in 2030 (not unlike how they misremember cooling predictions in the seventies). That's the danger of normalization. The Changes will ramp up our tolerance for barbarism while we bicker and dither.

There's a vision of this normalization in Kim Stanley Robinson's novel *New York 2140*. Set in its titular city and year, multimeter sea level rise overcomes a seawall and permanently inundates lower Manhattan. Instead of abandoning the city, people occupy buildings that occasionally collapse into eroded foundations. They travel by gondolas and skywalks, by boats instead of taxis. They trade financial instruments indexed to sea level rise and property values. All normalized. Robinson pitched the story as an absurd extension of capitalism beyond ecological limits, but it's not prediction. As Ursula K. Le Guin and Octavia Butler taught, science fiction isn't about the future; it's the now turned up a notch. What Robinson's vision of a flooded New York City suggests is how people might cope the day after disaster. We figure the coasts will be abandoned or ruined with seawalls. What if they aren't?

Elsewhere, there will be famine and war and rumors of war. Perhaps our descendants look back on the lack of urgency in the long twenty-first century as a great Dithering.[50] Maybe a planetary sovereign emerges along the lines of the Climate Leviathan or Climate Mao.[51] Or what if deadlines pass and, instead of the End Times, it turns out a Greenhouse Planet is negotiable so long as you have money?

Isn't that the lesson of Walter Benjamin's angel of history? In his famous ninth thesis on the philosophy of history, he writes:

> A Klee painting named "Angelus Novus" shows an angel looking as though he is about to move away from something he is fixedly contemplating. His eyes are staring, his mouth is open, his wings are spread. This is how one pictures the angel of history. His face is turned toward the past. Where we perceive a chain of events, he sees one single catastrophe which keeps piling wreckage and hurls it in front of his feet. The angel would like to stay, awaken the dead, and make whole what has been smashed. But a storm is blowing in from Paradise; it has got caught in his wings with such a violence that the angel can no longer close them. The storm irresistibly propels him into the future to which his back is turned, while the pile of debris before him grows skyward. This storm is what we call progress.

Unite a carbon atom with two oxygen atoms, you get energy. The side effect is a carbon dioxide molecule that inflicted mass extinctions in Earth's past. Energy is what history calls progress.

From prison during World War I, Rosa Luxemburg wrongly attributed a question to Engels that was, it seems, a misremembering of another's question or possibly just her own: "Bourgeois society stands at the crossroads, either transition to socialism or regression into barbarism." And if it was to be barbarism, at least workers might refuse the role of lackey to the ruling class and seize mastery

---

50   See Robinson, 2312.
51   See Wainwright and Mann, *Climate Leviathan*.

over their destinies. Not quite! Well, the stakes are higher now, yet we have less agency than ever. Can we alter course in a storm called progress? Let us reckon with our denial.

Socialism or barbarism? The Great War took the latter option. Then an armistice held for a few years until blood spilled again. After the Second World War's battlefields and genocide, major powers quieted down to proxy fights. The developing world suffered most. For decades, nuclear weapons nearly snuffed out humankind on multiple occasions. Life went on mostly oblivious. But in the midst of that nuclear danger, another was detected. President Johnson learned of it in a dull 1965 report. Exxon knew by the late seventies. The Charney Report in 1979. Hansen's Congressional testimony in 1988, same year as the IPCC formed. Report after report after report. Kyoto in '97, Copenhagen in '09, and Paris in '15. The public awakened but found few levers of power. Mass protests swelled and went ignored. Militaries armed themselves. Refugees were displaced. "You're not from around here, are you?" A possible Green New Deal, a decarbonization plan, net-zero commitments, fossil emissions still rise, grifts, and hopes, and barbarism. Hints of an energy transition. Barely a hope of socialism. What next? What next?

# Epilogue

Some years ago, my family and I were traversing the United States en route to the East Coast. I accepted a professorship in Baltimore and traded exposures to fires in mile-high Denver for new dangers near the harbor.

I have always loved the open road so dearly. Over thousands of miles, the landscape changes immensely. Much of middle America is flat and unremarkable space for crops. Road signs warn of God's judgment or the nearest gas station.

Southbound on a highway near the town of Tonkawa, off to my left, I saw oil wells in a vast field. All that is solid melts into air, so to speak. The area has been drilled and now fracked for its oil and gas over a century, a harm surpassed only by the violence inflicted for far longer against the Indigenous Tonkawa tribe. I didn't know the history. I perceived it narrowly, ignorantly as nondescript Oklahoma. Extract hydrocarbons out of the ground. Trap excess heat equal to nine hundred thousand atomic bombs daily. Pump all that heat into oceans as if there will be no repercussions. Oil wells decorate landscapes across United States. What caught my eye was off in the background, so distant they were barely visible through the haze. An array of wind turbines.

Oil wells in the foreground, wind turbines in the background. In the early days of this book, the scene struck me because it could be read two ways. In one interpretation, infrastructure neared retirement on depleted land, but beyond it stood the clean future. Or following the Jevons paradox, a profiteer maximized gains from the old fuels and the transition simultaneously. Either a vision of the Changes or a specter of the Dithering.

# Acknowledgments

This project was unexpected. Throughout 2019 and 2020 I felt frustrated as some of the most ambitious climate plans ever proposed on both sides of the Atlantic were blocked by electoral politics. Then a global pandemic threw physical safety, supply chains, and a dysfunctional society into question. Bold protests for racial justice that summer were met with violent repression from police and the right. A few months after the lockdowns began my first child was born.

I was a mix of joy and urgent concern when I awoke one morning to a message from Sebastian Budgen, who, having heard of my work from China Miéville and Richard Seymour, asked whether I might have any project suitable for Verso, a publisher that greatly shaped my education. I'm so deeply grateful for China, Richard, and Sebastian. Without them, this project wouldn't have begun. Conor O'Brien and Nick Walther were incredibly diligent editors, and I'm grateful to the whole team at Verso involved in production. My thanks also to Rosie Warren for including an early version of this material in *Salvage*, since her editing helped influence this book as well.

Thanks as well to Naomi Oreskes, Geoffrey Supran, Andreas Malm, Zeke Hausfather, Timothée Parrique, Lisa Schipper, and Neal T. Graham for corresponding on questions at various stages of my research.

I'm forever grateful for Deven, who believes in me, talks through these ideas when they are nascent and struggling to form, and encourages me always. Most of all to Logan and Asher, my reasons for writing about the future, I love you.

# Index

# Index

# Index